LORDS BUSHES

The History and Ecology of … Forest Woodland

M. W. HANSON

ESSEX NATURALIST number 7

Published by **The Essex Field Club,** c/o Passmore Edwards Museum,
Romford Road, London E15 4LZ
Edited by **David Corke** Ph.D., B.Sc., F.Z.S., F.R.E.S.
Illustrations by **M. W. Hanson**
ISSN 0071-1489 (for **Essex Naturalist** Series)
ISBN 0 905637 12 7 (for **Lords Bushes**)
Printed by **Plaistow Press Magazines Ltd**

CONTENTS

EDITORIAL

Since 1878, when Epping Forest was 'safeguarded for ever' by being given into the care of the Corporation of the City of London, members of the Essex Field Club have needed to defend the Forest against a long succession of threats. The threats have been of three main types: total neglect and lack of management reducing the diversity of wildlife, attempts to take the Forest lands to build railways and motorways and over-management turning parts of the forest into neatly manicured parkland.

An example of the latter type comes from a 1974 planning study by the Royal Institute of British Architects (61):

"By contrast Lords Bushes is enclosed, partly natural forest and partly artificial. The main drawback of difficulty of access for people with cars could be resolved by the provision of an informal car park within the grounds or behind the primary school site with access from Princes Road.

The existing artificial nature of Knighton Wood adjoining could be extended to Lords Bushes and amplified by planting additional non-indigenous trees and shrubs, by forming glades with fountains and waterfalls, and the introduction of a variety of water fowl to create a Regents Park atmosphere.

This would provide emphasis to the present contrast with the remainder of the forest and an area attractive to people with young children."

Fortunately, Lords Bushes has not yet been converted to an imitation of Regents Park. In this book, Mark Hanson explains the considerable historical and wildlife interest of Lords Bushes. Certainly this is the first time that one person has been able to give such a complete record of the flora and fauna of any part of Epping Forest.

Mark Hanson has shown that the accidental fires of 1976 have greatly improved the wildlife interest of Lords Bushes. If the Conservators read this book, they may well be able to further aid the wildlife of the area — but this time by deliberate action!

EDITORIAL ACKNOWLEDGEMENT

The Essex Field Club is extremely grateful to the Passmore Edwards Museum for making a substantial grant towards the cost of this publication.

Map 1. Lords Bushes and its environs as surveyed in 1772-4 by Chapman and André. Scale 1:20,000.

Map 2. Lords Bushes and its environs in 1979 (Ordnance Survey) Scale 1:20,000. (Reduced from 1:10,000 map with permission). Crown copyright reserved.

INTRODUCTION

Lords Bushes, an isolated portion of Epping Forest, is situated in the extreme south-west corner of the modern county of Essex in the parish of Buckhurst Hill (formerly Chigwell) just 15 km north-east of the City of London. Its grid reference is TQ4193 (46).

Until the drought year of 1976 some 97% of its 92 acres (37 ha) was covered by woodland, predominantly beech with oak, hornbeam and holly, many of the trees being pollards, relics of a wood-pasture system of management and characteristic of much of Epping Forest. Many trees, notably the beech, succumbed to the drought of 1976 and subsequently a large portion of the central area of the woodland was clear-felled of its dead and dying trees, leaving an extensive cleared area devoid of vegetation. Such clearings are rare events in ancient woodland today (almost invariably if a woodland is cleared today it involves its complete destruction i.e. the cleared area is turned over to agriculture or is replanted with conifers) and so I began recording the flora and fauna as it appeared on this area. An account of the results of this survey, and work carried out in other parts of Lords Bushes relatively unscathed by the events of 1976, and some historical research, are contained in the following pages.

WOOD-PASTURE

Wood-pasture and its relationship to other methods of woodland management

Before I go into detail about the history and ecology of Lords Bushes a brief note is required, to put it into its context as part of a wood-pasture system, and the relationship wood-pasture has to other types of woodland management. Much of this section is taken from Rackham (59).

Wooded areas in lowland Britain today are (or have been) basically managed in three ways — as plantations, woods or wood-pasture.

Plantations: A relatively recent innovation: in a plantation the trees are all of one age, usually of one species (commonly coniferous trees today) and are deliberately planted as a crop to be harvested at about a third of their natural life-span and the stumps left to die; the land either being re-planted or put to some other use.

The other two methods of management differ in that they are of greater antiquity and are more conservative in their treatment of the trees, which are in most cases indigenous to the site upon which the wood is situated. Woods and wood-pasture were certainly in operation in Anglo-Saxon times and woods may be of earlier origin.

Woods: In a wood the trees are basically of two types: timber (or standard) trees and underwood (or coppice). Periodically, on a rotational basis, a section of the underwood is felled: the poles being cut just above ground-level, from a stool, which then sprouts more shoots which eventually grow to provide poles for the next felling. The wood cut provides among other things tool handles, material for fencing (hurdles) and most often fuel. Interspersed among the underwood are the standard trees (usually oak); these are allowed to grow for many decades until a usable size is reached and are then felled and the timber used in building, for example, providing beams, posts and planks. This system of management is often referred to as 'coppice with standards'. The regrowth from coppice stools is much favoured by livestock and to prevent animals browsing woods are invariably surrounded by extensive earthworks (woodbank) supporting a hedge or fence.

Wood-pasture: This system differs from woods in that it combined the growing of trees with the grazing of animals. In order that this could take place the trees in a wood-pasture are pollarded (or lopped); in Epping Forest this was about 2-3m above ground level, leaving a permanent base called a bolling, which sprouts in the same way as a coppice stool but at a height at which livestock cannot browse the regrowth.

The word 'Forest' needs some clarification: in medieval times it was purely a legal term given to an area of land over which Forest Law had jurisdiction. The 'legal' forest could include farmland, villages, hamlets and even large towns like Waltham Abbey and Epping as well as (usually) being centred on a large tract of pre-existing common-land: the 'physical' forest which could include not only trees but grassy plains and heathland as well. It was only at a much later date that the word forest came to be applied to a large stand of trees.

Deer could roam unhindered throughout the legal Forest, commoners' livestock could roam the whole physical Forest irrespective of the manor from which they originated, but wood and timber could only be cut in the land owners' or commoners' particular manor.

Lords Bushes was part of the physical Forest in the royal Forest of Waltham. It was an integral part of the wood-pasture system which had been in operation probably since at least Anglo-Saxon times and until the accident of 19th century encroachment and subsequent urbanisation caused its separation from the rest of Epping Forest. Its predominant land-uses were grazing and pollarding carried out by the commoners and probably also the Lord of the manor of Chigwell, the manor in which Lords Bushes was situated.

PRE-DOMESDAY (1086) HISTORY OF THE EPPING FOREST AREA

The retreat of the Devensian glaciation about 10,000 B.P. (Before Present) and the corresponding amelioration of the climate over the next 2–3,000 years enabled the colonisation of the land by trees, culminating in what is known as 'wildwood'. The wildwood covered much of Britain before the emergence of man as a pastoralist, though it is thought that in some areas of Britain mesolithic hunters may have cleared quite appreciable areas of wildwood. From excavations carried out at High Beach it is clear they were present in what is now Epping Forest (7). It was, however, probably not until the late Neolithic (about 4,800 B.P.) and the Bronze Age in Britain that areas became consistently clear of natural vegetation as a result of settled agricultural practice, although Bronze Age remains are rare in Essex and such agriculture probably had little influence here.

In contrast there is much evidence of Iron Age settlement in S.W. Essex (Ambresbury Banks in Epping Forest is thought to be Iron Age in origin) and it is probable that at some time during this period land would have been cleared for agricultural purposes. The number of Roman 'finds' in the lower Roding Valley; for example, the Roman road, the villa at Woolston and the many lesser finds of Roman material in the Chigwell area, indicates that this part of the Roding Valley was at least partly open, settled and cultivated land by late Roman times (14).

Place names such as Loughton, Leyton and Waltham indicate the Epping Forest area was not of late colonisation by the Anglo-Saxons, they may well have settled on the surviving Roman agricultural landscape. The name Buckhurst Hill is thought to be of Anglo-Saxon origin derived from Boch=beech and hyrst=grove (14).

With the coming of the manorial system, Lords Bushes must have had its first associations with what was to become the manor of Chigwell (with which I assume it was subsequently associated until the Epping Forest Act of 1878). Squirrels Lane (see map 3(b)) which connected Lords Bushes with Chigwell, could well have originated at this time as the routeway by which the inhabitants of Chigwell manor could reach their wood-pasture to graze their animals or cut their firewood. The service trees which border what remains of this routeway (see page 39) may be an indication of its antiquity.

The Domesday survey of 1086 shows a very well established agricultural landscape in the Epping Forest area (14), probably the Forest itself being managed as wood-pasture and the more fertile soils of the Roding Valley being used for arable crops and probably also some pasture as well.

HISTORY 1086-1700

There is sadly scant information in the literature concerning Lords Bushes for this period; presumably the manorial roll which might have yielded information no longer exists. Lords Bushes being considered 'commonland' (as opposed to privately owned coppiced woodland) may well have been relatively poorly documented in any case. There are, though, a number of references to the land surrounding Lords Bushes and these give us some insight into the history of the area. It is, however, well to remember that over 600 years of history has been compressed into a few paragraphs.

The first mention I can find is in 1135 when William de Montfitchet granted by charter to the Cistercian Abbey of Stratford Langthorne "all his lordship of Ham, 11 acres of meadow, two mills by the causeway of Stratford, his wood of Bochherste and his tithe of pannage" (60). The Bochherste referred to is probably a woodland that existed, in part until the early 19th century south of Lords Bushes and abutting Pea-field Wood (now built on) and part of Knighton Wood (23). The wood was subsequently recorded under a plethora of names, Monkene Bukhurst (1286) and Buckhurst alias Monkyn (1521) among them, finally it seems it was called Monkham Grove (60). Pannage refers to the practice of fattening swine on the autumn beech mast and acorns, pigs were an important source of meat in medieval times and it is possible this practice would have occurred in Lords Bushes. Monkham Grove, in contrast to Lords Bushes, was managed as a coppice wood, probably from a small dwelling house established by the Abbey nearby. In 1514 the Abbot of Stratford leased to John Crofte "citizen and grocer of London all the wood and underwood, ditches and hedges and a sufficient dwelling house thereon . . . at 40 shillings yearly rent" interestingly there was also a proviso that the lessee should always keep twenty "stadils" (young trees) to the acre and to keep the grove enclosed (60).

In 1360 Edward III through his son John of Gaunt bought a messuage and 92 acres of land (including 20 acres of meadow) (71). This probably refers to land and buildings that were situated at the junction of what are now Westbury Lane and Kings Avenue. It seems one or all of the buildings became royal establishments (hence Kings Avenue), one was probably a hunting lodge: a lodge is marked on the Chapman and André map of 1777 (see page 6). If it was a hunting lodge it was unusual in being situated outside the physical forest, unlike Queen Elizabeth's Hunting Lodge and the New Lodge extant until the 1890's on Fairmead Bottom (24). A Swainmote, one of the higher Forest Courts, was held at Bukkershyl in 1495 (26) and its most likely location would have been this building. It was probably the close proximity of this royal establishment that gave its name to Lords Bushes, which became known as Lodge Bushes, this name appears on a map of the keepers walks in Waltham Forest drawn up in about 1641 (72). 'Lords' appears to be a 19th century corruption perhaps resulting from a word-of-mouth error or error in transcription. It appears in one 19th century document (the deeds to a house) as "Lords otherwise Lodge Bushes", though William Cole, a local naturalist, as late as the 1880s still referred to it as Lodge Bushes. I should also add that its name appears not to have been well known, Richard Warner the celebrated Woodford botanist referred to it merely as the "wood between Muncombe and the Bald Faced Stag" (74).

In 1572 attempts were made to enclose the adjacent Knighton Wood (then, as Lords Bushes, wood-pasture subject to the rights of the commoners) which led to violent demonstrations by the commoners. There were more attempts in 1670 and 1781 (14), the latter presumably being partially successful and forming the basis of what was to become the Knighton Estate, extant until 1928 and latterly the home of the Buxton family.

Lords Bushes throughout this period seems to have had a rather undistinguished history, probably the people of Chigwell manor just getting on with their tasks of pollarding and depasturing their animals as centuries old tradition demanded.

HISTORY 1700-1900

The eighteenth and nineteenth centuries were a period of mixed fortune for Lords Bushes, as indeed they were for the rest of Epping Forest. The 19th century in particular saw the declining interest of the Crown in the deer of Epping Forest, the red deer disappeared and the fallow had declined almost to extinction by 1870 and illegal enclosure, was widespread thoughout the Forest. Grazing and wood-cutting by the commoners still continued and it was the attempted suppression of the commoners' rights, through illegal enclosure,that caused the well-known events that led up to the Epping Forest Act of 1878 (1).

Forest law during this period, particularly in the 19th century, seems to have been rather slackly administered and illegal enclosure was not the only abuse recorded. In 1813 the Warden of the Forest, Mr Long Wellesley, consulted his forest officers as to the nature of the prevailing abuses; they included illicit gravel extractions, turf removal, felling of pollards, and deer poaching (26). When detected, these offences against the vert and venison were presented before the court of attachment of the Royal Forest of Waltham (held at the Kings Head, Chigwell until 1843) which was presided over by verderers. Though having only extremely limited powers, its purpose was merely to enquire of offences, but in a few cases it could order an enclosure to be taken down, or confiscate weapons used in poaching. Conviction and sentence was left to the higher courts (the swainmote and justice seat: the last justice seat appears to have been held for Waltham Forest in 1670!) (26). The court records, the roll of the court of attachment, record the presentments before the court and obliquely (through the minor offences and occasional grant of a licence of one sort or another) tell us much about Lords Bushes at this time (22).

From the mid-eighteenth century onwards a noticeable feature of the court records is the number of licences granted, usually to the big landowners to "hunt, hawk, course, sett and shoot" with of course the stipulation that the King's deer (red and fallow are mentioned) were the excepted quarry. Remembering that Lords Bushes at this time were situated in what was very much a rural area (see map 1.) it is certain that deer (probably fallow deer) would have been present in Lords Bushes. In the mid-eighteenth century Thomas North and his wife Mary, who resided at Monkham House, were presented before the court of attachment for letting two brace of greyhounds chase the deer and for "erecting a high pale round the said grove [Monkham] whereby the deer cannot have free passage" (60). The Norths were also presented before the court on several occasions between 1737 and 1757 for illicit felling in Monkham Grove (60), at that time and until the end of the 18th century, licences were required even for such routine management of woods that were outside but very close to the physical Forest (Rackham pers. comm.).

In 1809 and again in 1810 and 1813 George Brühl (son of the Ambassador Extraordinary of Saxony) who farmed Kings Place farm was presented before the court for hedging and ditching across Kings Place Lane, near Lords Bushes, obstructing the then well used lane. This lane in one direction connected with a route to Woodford and in the other (via Squirrels Lane, over a ford on the Roding and up Luxborough Lane) led to Chigwell and Woodford Bridge (2) and was then a main thoroughfare (the B170 Roding Lane used today was not built until 1890) (71). The court roll states the lane had been "free to passengers time out of mind" and the fact that neither horse nor cart could pass the obstruction must have been a great inconvenience to local people, and as the court pointed out was hindering the keepers in the execution of their duty, but they had to agree that it was in no way infringing forest law, and reluctantly they could make no order. Brühl, a few months after his second presentment for hedging and ditching, was presented for felling an oak pollard and four other trees adjoining his land on the edge of Lords Bushes, but apparently again no order was made (22).

The most insidious threat at this time, not only to Lords Bushes but to the whole of Epping Forest, were the enclosures, and presentments for this offence figure heavily in the court records in the nineteenth century (22). The courts themselves virtually surrendered to the enclosures by passing a resolution that they would not interfere with small encroachments on the sides of public roads, or lanes, or in any other places within the legal boundaries of the Forest where the deer did not usually depasture, but that encroachments without the consent of the Lord of the manor or where deer usually depastured were to be thrown open. Lords Bushes in the eighteenth and nineteenth century illustrates well this relaxed attitude towards enclosure.

In 1779, Thomas Nicholson, publican of the Bald Faced Stag, was presented before the court for enclosing a piece of ground from the Forest (probably near the junction of what is now Knighton Lane and the High Road) and he was ordered to throw down the enclosure by the court. Obviously he failed to do so for six years later in 1785 and again in 1786 he was ordered to throw down the enclosure, finally doing so on his fourth presentment late in 1786; but a further complication had arisen for Thomas Priest, a blacksmith, had built his Smithy Shop on the site enclosed by Nicholson and he in 1787 was ordered to take down the Smiths shop. Yet again the limited power of the court is illustrated by Thomas Priest (still a blacksmith) appearing in court in 1800 where it is recorded that he "do continue the inclosure of ground" this time 60 rods of land from Lords Bushes upon which he had built a cottage, nothing obviously came of further orders to disenclose for in 1803 the court decided thet Priest could lawfully continue the enclosure of ground, the enclosures not being considered "prejudicial to his majesties vert and venison" (22).

There were other encroachments in Lords Bushes at this time, mainly in the north west corner. Land now occupied by houses, bungalows, the ambulance station, the cottage (which was to become the cottage hospital in 1868), Buckhurst Hill House, Holly House (now a private hospital) and Knighton Lane itself are all built on land that once was part of Lords Bushes (further evidence of this is provided by the oak and hornbean pollards which protrude from the pavement along Knighton Lane!).

In 1843, John Hatherhill, Keeper of the Loughton Walk, recorded numerous minor enclosures around the Bald Faced Stag but also more substantial ones of 5, 2½ and 1 acres (22). Overall probably in excess of 16 acres were lost to the enclosures from Lords Bushes, we know this because today, its area is approximately 92 acres, while in 1797 the property of James Hatch of Chigwell was assessed (22), including Lords (then Lodge) Bushes and was found to have an area of 108 acres — assuming the same unit of measurement was used (and used accurately) it thus seems likely that at least 16 acres were lost.

The Buckhurst Hill we see today, centred on Queens Road, was farmland in the early 1800's, with just the Bald Faced Stag, Kings Place, Monkhams Farm and the odd cottage as the only habitations. The extension of the Great Eastern Railway from Woodford to Loughton in 1856 preceded the rapid development of this area as a London suburb (71). First Queens and Princes Roads were developed but it was not until the 1930's that Forest Edge, Broadfield Way and part of the Knighton Estate were built on, effectively making Lords Bushes and Knighton Wood a suburban island of woodland.

The house I an now writing this essay in was part of the mid-Victorian development of Buckhurst Hill; built probably in 1864 on land enclosed from Lords Bushes by the Lord of the manor of Chigwell, James Mills, who had "purchased" the Forestal rights from the Crown in 1858 (71). The land was "enclosed with other parts of such waste [i.e. Lords Bushes] by said J. Mills as Lord of the said manor with the consent of the Homage according to the Custom of said manor", so reads the Abstract of title. James Mills had probably instituted the manorial Court Customary in order to make it appear the commoners were agreeing to the enclosure and subsequent loss of common rights in Lords Bushes. The land was sold to Ephraim Salter (tenant of the Bald Faced Stag since 1850, he also operated the Buckhurst Hill Brickworks in Lower Queens Road) for £46 and he presumably had the terraced houses built (now 160-166 Princes Road) for in 1865 he sold the plot for £580! Interestingly Ephraim Salter bought the freehold of the Bald Faced Stag from James Mills in 1868 (22).

The matter did not end there, for the original enclosure was deemed to be illegal and the arbitrator, Sir Arthur Hobhouse, following the Epping Forest Act of 1878, ordered the then owner of the house, Richard Perkins, to pay the sum of £15.6s.8d. to the conservators and upon such payment "the said hereditaments should be quieted in title against and be released from all right of common and to be held in all respects as if the enclosure thereof had been a valid and proper enclosure."

In 1872 another railway was built, this time an experimental one (sometimes referred to as a tramway (see Plate 10)) constructed actually in Lords Bushes, apparently down Monkams Lane where 1700ft of track was laid consisting of two parallel level faced timber baulks with a central steel guide rail. The railway was built in Lords Bushes to test a railway system destined for the Lisbon Steam Railway Company in Portugal (35).

1878 saw the passing of the Epping Forest Act and Lords Bushes became part of Epping Forest, its boundaries now fixed precluding any further possibilty of enclosure. The Forest as a whole at this time was well used as a recreation area, facilitated by the extension of the railway. The royal visit by Queen Victoria in 1882 to dedicate the Forest, finally settled its future as a public recreation area. Concomitantly, a need for refreshment areas for the thirsty and hungry visitor arose and so it was that the famous Epping Forest 'retreats' came into being (73). As early as 1882 there was the 'Forest Retreat' in Princes Road, backing on to Lords Bushes and run by Jeremiah Bushell. The retreat had seating for 500 (perhaps an indication of the degree of use as a recreation area Lords Bushes was subjected to at this time) and donkey rides and swings were available as well. The retreat was run later by a member of the Riggs family and was used as a place of refreshment up to 1935, after which it became a pigeon fanciers' club and is currently renovated as a private dwelling.

In 1883 Professor G. S. Boulger, president of the Essex Field Club, in the company of E. N. Buxton, made a "somewhat careful examination" of Epping Forest starting off from Lords Bushes, in order to assess the action required for the Forest's future management (6). Buxton lived at Knighton, the estate then adjacent to Lords Bushes, of which the 37 acre Knighton Wood is all that remains, given to the Forest by the Buxton family in 1930. Buxton and Boulger noted that Lords Bushes was already being thinned of its "meaner" hornbeams while the less "hopelessly deformed" pollards, it was suggested, be allowed to spread laterally by re-pollarding, though Boulger added that great vigilance would be needed owing to the increased possibility of illicit lopping (lopping was declared illegal by the Epping Forest Act of 1878). Lords Bushes was thinned out four times in the 14 years after 1880 and particular note was made of the wealth of undergrowth and luxuriance of the colonising birch seedlings (then an uncommon tree in the Forest). William Cole, writing in 1894, said of Lords Bushes, "In our own knowledge from being twenty years ago a dark dismal place doomed to the clutches of the speculative builder, with an experimental tramway and a broad road to be bordered with 'eligible modern villas', Lords Bushes as now transformed is one of the most beautiful and luxuriant woods in the Forest, and this in spite of its nearness to a large village and railway bringing down crowds of excursionists and school children" (13).

Boulger also noted the effects of gravel digging in Lords Bushes, today the former gravel workings are very much in evidence in the north west corner (see map 3(b)) and though made illegal by the Epping Forest Act (33) the parish surveyors were noted extracting gravel in 1883 (6), obviously still having the right to do so. Gravel extraction on a large scale may well have been a nineteenth century phenomenon in Lords Bushes. Lords Bushes may have been one of the few places locally where there were easily worked surface deposits of gravel. Elizabeth French the founder of the now international firm French-Kier started her business in Buckhurst Hill, carting gravel, about 1860 (71); also E. N. Buxton had Knighton Pond dug out at about this time and was able to sell the extracted gravel. Gravel was probably used for road repair and construction.

Boulger did note that many of the trees around the excavations were being undercut (it being illegal to fell the trees and this undercutting is still visible today) and he also adds that the diggings could be "natural, irregular and picturesque pieces of water, instead of being elaborately drained as at present" (6) this could well account for what were until recently the two good ponds (see maps 3(c), P1 and P2). Buxton himself may have superintended the construction of the concrete dam (middle background of Plate 4) to create a pond; today, sadly, the ponds are infilled with beech leaves from the surrounding trees and the dam has been broken and thus the ponds have only a poor flora and fauna.

HISTORY 1900—PRESENT DAY

After a much troubled nineteenth century, Lords Bushes entered the twentieth century relatively secure in its boundaries, no longer exploited as a natural resource for wood, pasture or gravel but as a popular recreation area for London's public. Events since then have been rather unremarkable. The woodland suffered from fires in 1921 (though not to the extent of the 1976 fires). The 1930's brought increasing urbanisation to the Buckhurst Hill area, in 1936 Monkhams Farm was sold and demolished (60) and this was followed by the development along what is now Forest Edge and Broadfield Way. About this time the 'retreat' in Princes Road ceased to operate, indicative of a decline in the number of visitors to this part of the Forest. Following the development in the late 1930's the woodland was cleared of its undergrowth and small trees along the Forest Edge side (Scott pers. comm.). During World War II, two bombs (including a parachute bomb) fell on Lords Bushes creating the two craters we see today (see map 3(b)).

The prevailing drought conditions of 1976 drastically altered the ecological character of Lords Bushes (see section on the flora). Today the wood is used mainly by local people walking their dogs, though in spring and summer non-local people visit Lords Bushes, though perhaps more frequently the adjacent Knighton Wood. Other users include 'treasure hunters' who use metal detectors to discover buried artefacts, orienteers and also (though very rarely now since the move of the Nightingale Riding Stables from Buckhurst Hill High Road) the occasional horse rider. In December to February 1979/80 holly and further dead beech trees were cleared from Lords Bushes by the Conservators (see Plate 1) due to concern about further outbreaks of fire and undesirable characters lurking in the bushes.

PHYSICAL FEATURES

Geology

The two dominant features of the geology of Lords Bushes are:

(1) Eocene London Clay

(2) Pleistocene Sands and Gravels.

Over much of Lords Bushes the London Clay is overlain by an outlier of sands and gravels presumed to be glacial in origin, London Clay outcrops only along the Forest Edge side of Lords Bushes (see map 3(a)). The sands and gravels are a very variable deposit in depth, being deepest in the north west corner of Lords Bushes (notably where

Map 3.
(a) Geology
(b) Surface Features
(c) Recording Units

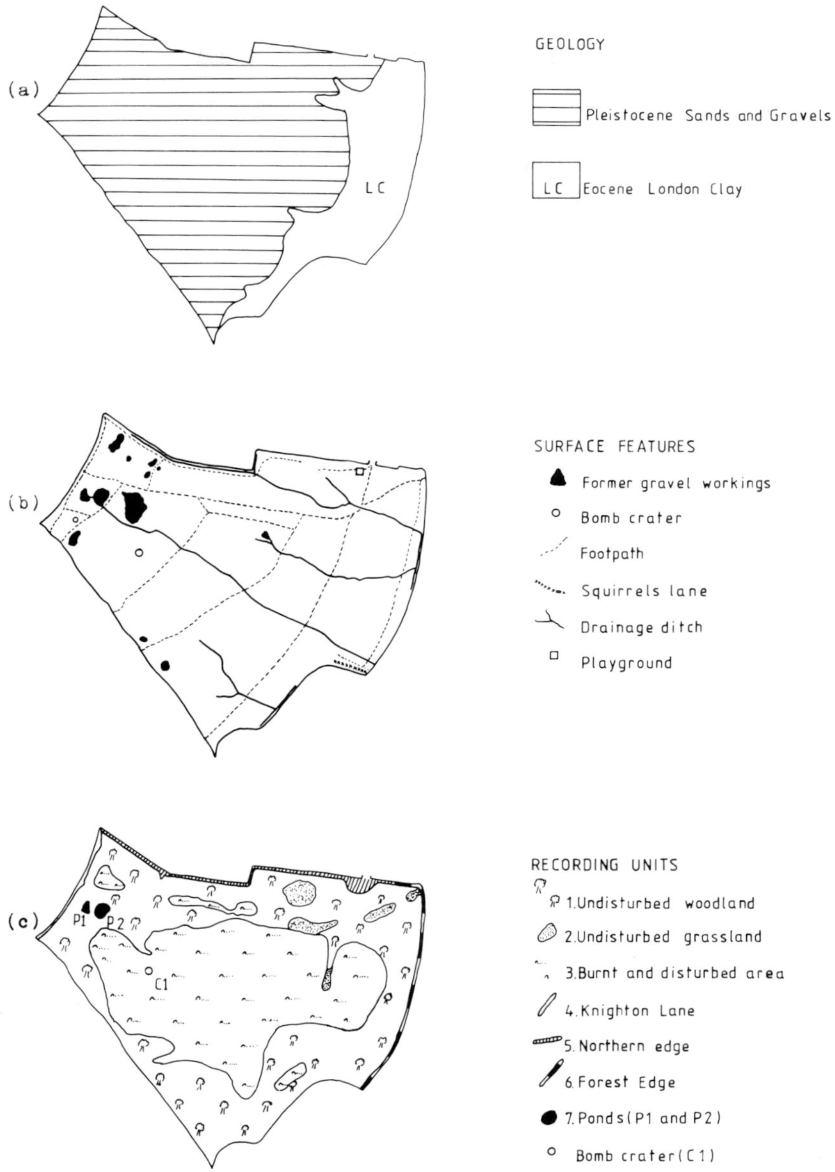

most of the former gravel workings are to be found (see map 3(b)). Here the deposit is probably well over 2m thick in places and in general the depth of the deposit decreases downslope to Forest Edge where the London Clay outcrops. The decrease in depth is not uniform and, from studying material caught up in the root systems of fallen trees, it is clear that London Clay can be near the surface in areas where there are apparently thick deposits of sand and gravel. The map of the geology, also drawn from evidence of material caught up in the root systems of fallen trees, shows approximately the extent of the sands and gravels to a depth of a few centimetres (though it should be noted that on a national level superficial deposits such as these are mapped only if they attain a depth of 1m or more). A further point to note is that the division between the two deposits is not as marked as is shown on the map and physical processes such as downslope erosion must further blur the boundary between the two formations. London Clay can also be exposed in areas of sand and gravel by the erosion of ditch sides.

Topography and Aspect

Lords Bushes is on a south-east facing slope overlooking the valley of the River Roding. The land surface drops fairly evenly over a distance of 825m (a slope of 1:20) from a maximum height of approximately 79m (O.D.) in the extreme north west corner to a height of about 38m (O.D.) on the eastern edge (Forest Edge).

Drainage

Drainage is achieved by four main ditches (see map 3(b)) at least two of which were probably dug in order to drain gravel workings. One of these (the longest) was re-excavated in early 1979

Soils

Woodland soils are very complex and a number of factors such as the underlying geology, slope, drainage and nutrient status must all be taken into account when assessing the soil types of a given area. In Lords Bushes soils on glacial gravels are acid, base deficient and a characteristic feature is a relatively deep layer of partly decomposed leaf-litter: a 'mor' almost peaty in consistency.

The climatic regime of south-eastern England produces wet conditions in winter, while the soils tend to dry out in the summer (spectacularly so in 1976).

Surface features

Lords Bushes being formerly part of a wood-pasture system lacks the woodbank typical of ancient coppiced woodland. Around much of the perimeter of Lords Bushes are ditches, presumably of recent origin to define the boundary of the woodland against possible encroachment by the adjacent houses, as well as for drainage purposes. Those dug on the border of Monkhams Lane prevented cars from entering on to the woodland. Until quite recently vehicles were able to go up and down Monkhams Lane, access is now restricted by a series of wooden posts across either end of the lane.

Knighton Wood, on the other side of Monkhams Lane, has a woodbank:

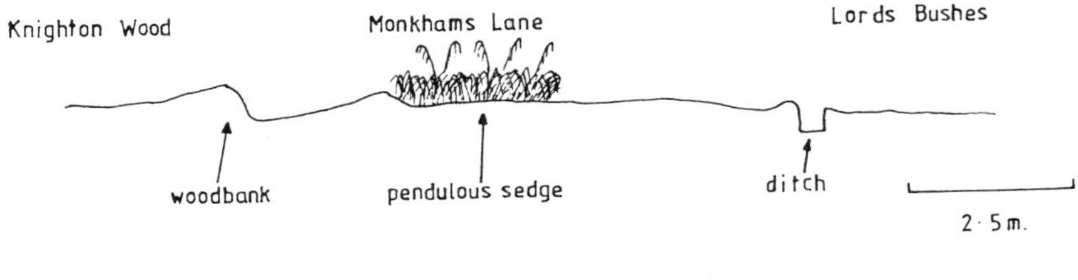

This feature presumably resulted from one or other of the three attempts at enclosing the wood (in 1572, 1670 and 1781); prior to this Knighton Wood itself was wood-pasture, as evidenced by a number of fine beech pollards, and would not have required a woodbank. The present ditch and bank is very irregular in size, in some sections it is barely perceptible and yet in others a double bank appears to be present: nowhere is the woodbank very large, a further indication of its relatively recent construction. Along the top of the bank are to be found a number of coppiced beech and hornbeam trees as well as standard oaks and at one point a rather poorly pollarded wild service. Pine and rhododendron have obviously been planted on the bank.

Monkhams Lane itself has been colonised at one point by sallow and in two places quite extensive areas of pendulous sedge.

Apart from gravel workings and their associated drainage ditches, Lords Bushes is apparently devoid of any obvious recognisable pre-twentieth century man-made feature. There are, however, three holes about four feet square and rather shallow, these may have been trial gravel diggings or possibly one of these is the "coal hearth" (used in charcoal burning) which is mentioned for Lords Bushes in the Victoria County History of Essex (70). On the north edge of the wood are the remains of a small 'dump' from one of the cottages that backed on to the wood.

THE FLORA OF LORDS BUSHES

Introduction

In order to put the present day flora of Lords Bushes into historical perspective I have felt it necessary to include brief notes based on work carried out elsewhere in Epping Forest and summarised mainly from Corke (14). These notes deal with the flora of the prehistoric wildwood, its relationship to the present flora, and most importantly, how management as wood-pasture has affected the flora. Further sections include an appraisal of the few known historical records of plants for Lords Bushes and also a detailed account of the present day flora including bryophytes and lichens of Lords Bushes.

The Wildwood

Recent analysis of the pollen record from bog deposits has shown that the prehistoric wildwood in lowland England was dominated by lime (*Tilia cordata*) and recent studies suggest this is true also for what was to become Epping Forest (3). A characteristic feature of the pollen diagrams from all sites in Britain is a decline, often rapid, in the frequency of lime pollen at certain points formerly thought to be climatic in origin. This decline is now attributed to clearances by man. Lime decline was not a synchronous event, it took place in various areas at different times but mainly in the Neolithic, Bronze and Iron ages. The lime decline in Epping Forest was remarkably late, radiocarbon dates indicate the middle Saxon period (as late as 950 A.D.) and was possibly associated with the rise of wood-pasture.

In the Epping Forest area lime gave way initially to birch and ultimately to hornbeam and beech. Oak throughout this period maintained a fairly constant presence in the pollen diagram, increasing a little after the lime decline. These species are characteristic of the Epping Forest area (including Lords Bushes) today.

Lime is no longer present as an indigenous tree in Epping Forest (it is present as a rare planted tree, there are two by St Johns Pond, Buckhurst Hill). It is thought lime was unable to maintain a hold because of the rise of the wood-pasture system in Epping Forest. Lime is sensitive to browsing, it only survives as a native tree in Essex in coppice woods (59).

Tree Species

It is very likely that Lords Bushes (as part of the wildwood) would have been predominantly lime. The abundance of beech and hornbeam we see today is probably only of recent origin surviving along with oak (specifically *Querus robur*) because they are unpalatable to browsing livestock. Grazing animals are the probable reason for the virtual absence from Epping Forest of many otherwise common, but palatable, tree species among them hazel and ash. Ash occurs as a single tree in Lords Bushes which was probably able to establish itself in the late nineteenth century following the cessation of grazing. Hazel grows very well locally, outside the Forest. I have a rather prolific coppiced example in my garden which is near Lords Bushes; it was also recorded along with ash growing in a hedgerow on Squirrels Lane (now sadly built on) just outside Lords Bushes (25). There it was probably a long established hedgerow tree free from the pressure of browsing in the nearby physical Forest.

The Plains

In wood-pasture there is a conflict of interest between the grazing animals and the trees — the more trees the less grazing and the more animals, the more difficult it becomes to replace the trees. In Epping Forest this conflict was resolved by having definite grassy plains and heathy areas where grazing was concentrated and by pollarding the trees, which if persistently managed in this way, grow indefinitely (standard trees have a definite life-span).

Although prior to 1976 Lords Bushes was almost 100% dense woodland this cannot always have been the case. The name 'Lords Bushes' implied some degree of open 'shrubby' growth (Rackham pers. Comm.) and the pollards indicate that animals were grazed there. Notable also is the presence of four small patches of grassland still today called plains. These patches of grassland are the remains of much larger open areas and as I shall argue later were much more of a healthy nature than is apparent today. The plains existed until a hundred or hundred and fifty years ago when grazing became slack and was finally terminated by urbanisation, since when they have suffered from a continuous encroachment by secondary growth.

HISTORY OF THE FLORA 1771-1974

There are no early complete records of the flora of Lords Bushes, though the Woodford, Roding Valley and Luxborough Lane area was favoured by the eminent local botanist Richard Warner author of the famous 'Plantae Woodfordienses' published in 1771, one of the first comprehensive local floras ever produced (74). This little book now makes rather melancholy reading, the decline and indeed outright loss of many plant species (due mainly to habitat destruction, and lack of habitat management) in the areas that Warner botanised in, is very marked.

Warner recorded species such as primrose (*Primula vulgaris*), common cow wheat (*Melampyrum pratense*) and slender St Johns wort (*Hypericum pulchrum*) as being very common in local woodlands and though all three species are still present locally they are either extremely rare or very uncommon. In additon to these generalised records. Warner makes a number of interesting specific references to plants found in or about Lords Bushes, among them mistletoe (*Viscum album*) which he recorded on oak, stinking hellebore (*Helleborus foetidus*) which he found "opposite a farm house at Muncombe" (probably a garden escape) and most interestingly the orchid autumn ladies tresses (*Spiranthes spiralis*) "found on the Forest near the Bald Faced Stag on the right hand of the road, not far from the eight mile stone." (Note: the public house known today as the Bald Faced Stag is labelled as the Bald Stag on the Chapman and André Map of 1777).

Warner's contemporary Edward Forster, a city merchant and amateur naturalist, had three sons and a daughter, the sons all taking an interest in botany and the father provided each with a copy of 'Plantae Woodfordienses'. Benjamin Meggott Forster's annotations to this work provide us with the record of "*Serapias latifolia* [more familiarly known today as *Epipactis helleborine* the broad leaved helleborine] found on a hill called the Hawk nr Chingford Green, also near the Bald Faced Stag in the wood there" (65). Sadly this record is of historical interest only, this orchid no longer occurs in the now built up area of Buckhurst Hill. Benjamin's brother Edward was perhaps the best known of the Forster brothers (giving his name to the woodrush *Luzula forsteri*), he found the now nationally rare all-seed (*Radiola linoides*) near the Bald Faced Stag (66) (in Essex now found only on Tiptree Heath) and also provided the only record of the moss *Antitrichia curtipendula* for Epping Forest "between the Bald Faced Stag and Monkhams" (34). This could well be the first bryophyte record for Lords Bushes, but sadly this moss is no longer to be found in Epping Forest nor indeed in the county of Essex. Other historical records of interest, because they are of species typically found on heaths and bogs, include heath woodrush (*Luzula multiflora*) and ivy-leaved bell-flower (*Wahlenbergia hederacea*) recorded in Gibson (29) as being found near the Bald Faced Stag. Pettifer (55) records the finding of the moss *Sphagnum plumulosum* near the Bald Faced Stag by Edward Forster in 1800.

With the formation of the Essex Field Club in 1880 a number of excursions were made to the Lords Bushes area. An early field trip describes the flora of the now virtually non-existent Squirrels Lane; this extract is from a field excursion made in July 1882 (25):

"The guides led the way through the pretty isolated portion of Epping Forest known as "Lords" or "Lodge" Bushes and so into a lane (commonly, but we understand from Mr D'Oyley, erroneously called "Squirrel's Lane") refreshingly rough, untrimmed and unimproved, with tangled grass and sedge, amid which many wild flowers nestled and hid their modest charms. Betony and Woundworts, Self-heal (*Prunella*), Skull-cap, Veronica, plenty of Trefoils, Clovers and Vetches, flourished everywhere whilst with more vigorous growth the giant Umbelliferae thrust themselves into notice, the branching heads of small white flowers contrasting well with the dark green hedgerows of maple, nut and ash, in which the Black and Red Bryony, Convolvulus and Bed-straws lovingly twined and blossomed. Here were noticed *Vicia hirsuta* and the delicate little Vetch, *Vicia tetrasperma*, many of the pods of which contained only three seeds, provokingly defiant of its learned name. Also many splendid plants of the ever-welcome St John's wort (*Hypericum perforatum*), the slender woodland beauty (*H. pulchrum*) having been already noticed in Lords Bushes; and to crown the whole, tall plants of Meadow-sweet waved masses of yellowish white flowers above their fellows and gave the charm of odour."

A year later following a trip aimed at producing management plans for the whole of Epping Forest, Professor G. S. Boulger in the company of E. N. Buxton gave the following description of Lords Bushes (6):

"Entering it (Epping Forest) at Lords Bushes, close to Buckhurst Hill Station, we had an example of the felling of Pollards that has already been done. Though not here, perhaps, excessive in the proportion of trees removed, there was not much judgement evinced in their selection.

There is here a valuable and picturesque undergrowth largely composed of Holly, which being evergreen, flourishes under shade too dense for the Hawthorn, and might well be planted in many other parts of the Forest. It was generally agreed among our party that the underwood should be preserved in every part of the Forest, if only for the sake of the birds; and that, though it was desirable to gradually eliminate the most decayed and hopelessly-deformed Hornbeams, no pollard Oaks, and perhaps hardly any Beech need be felled, these species making fine trees even after pollarding."

It seems that by the late 19th century there was little in the flora that excited the botanists of the day and even then it appears that the woodland was becoming overshadowed and a rather misguided management policy was attempting to remove some of the less aesthetically pleasing pollard hornbeams (Plate 8). Beyond records in 1899/1900 of a fungal disease attacking birch trees in Lords Bushes (probably the microscopic fungus *Melanconis stilbostoma* (51), I can find no further records until the late 1930's, when the woodland along Forest Edge was cleared of its undergrowth and small trees (this explains the number of apparently recently coppiced trees along this edge of Lords Bushes) for a distance of some 30 yards into the woodland. As a result of this some plants including figwort (*Schrophularia nodosa*) and common cow-wheat became common for a short period of time (Scott pers.comm). More recently the only specific higher plant record I can find in the Flora of Essex (34) is of the black sedge (*Carex nigra*). Thus it appears that Lords Bushes has received scant attention from botanists in the past and following the drought and clearance of 1976 with the subsequent appearance of a number of interesting plant species I decided to produce a flora for the whole woodland.

THE PRESENT DAY FLORA OF LORDS BUSHES

Recording for the flora was carried out mainly between August 1977 and August 1979. For convenience I have divided Lords Bushes into 7 fairly discrete recording units (see map 3(c)), and a brief description of these areas is given below.

(1) The Undisturbed Woodland

This recording unit embraces all that remains of the woodland that covered some 97% of Lords Bushes prior to 1976 and included what is now the burnt and disturbed area (recording unit (3)) and it is useful to remember this when considering the flora of the burnt and disturbed area.

The structure of the woodland, on the sands and gravels, consists typically of an interlocking canopy of beech trees with, in places, a dense shrub layer of holly and the occasional bramble (*Rubus sp.*) and other species of tree such as oak, hornbeam, or birch and very little else, the field layer is virtually non-existent. On the London Clay too there is a dense stand of trees consisting of oak, hornbeams and lesser numbers of other tree species such as field maple and wild service and similarly there is a virtually non-existent ground flora.

In recording the flora of the undisturbed woodland I have largely ignored 'incidental' species, mostly weed species such as groundsel (*Senecio vulgaris*) and fireweed (*Epilobium angustifolium*), that have colonised bare grund for example where a pollard has been blown over. I feel that to record these would be misleading, because the ground flora of the undisturbed woodland is so poor. I have, though, included native woodland species such as blue-bell (*Endymion non-scriptus*) enchanter's nightshade (*Circaea lutetiana*) foxglove (*Digitalis purpurea*), and remote sedge (*Carex remota*) which although present are in extremely low numbers and found only in places where some light is able to reach the woodland floor, they are not representative of the overall nature of the woodland, but they may well be relics of the flora that occurred in this woodland when actively pollarded and before beech came to overshade the woodland. I should also mention that a number of tree species such as Swedish whitebeam (*Sorbus intermedia*) ash (*Fraxinus excelsior*) yew (*Taxus baccata*) and sessile oak (*Quercus petraea*) are present only as individuals or a few trees and some species, such as Swedish whitebeam, are non-native.

(2) Undisturbed Grassland

There are small patches of long-established grassland, remnants of much larger areas, that have been encroached upon by secondary growth since the cessation of

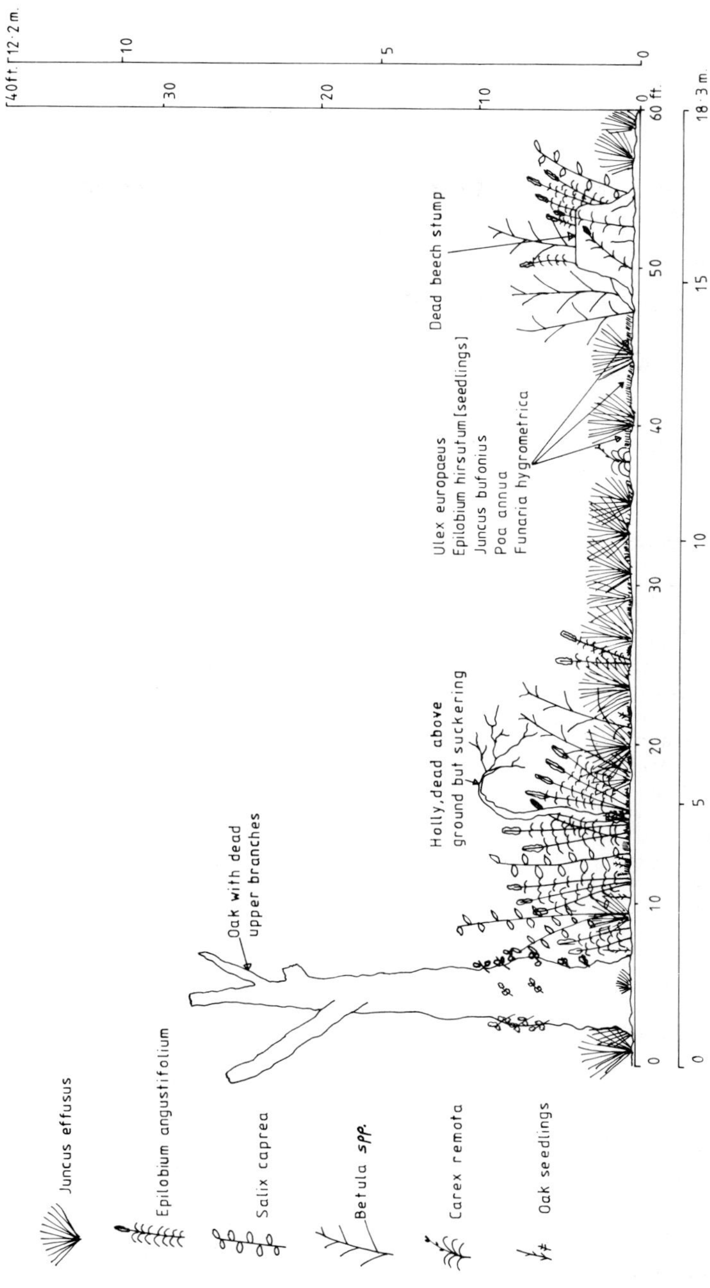

Simplified structure of regeneration of the burnt and disturbed area on glacial gravel in Lords Bushes (October 1978).

grazing activities in the nineteenth century. Although small in size they are still referred to as plains with picturesque names such as 'Daisy' and 'Jericho'. The flora of these grasslands today include 'heathy' species such as sheeps sorrel (*Rumex acetosella*), heath bedstraw (*Galium saxatile*) and heath grass (*Sieglingia decumbens*).

(3) Burnt and Disturbed Area

Lords Bushes, prior to 1976, was for the most part mature deciduous woodland dominated by beech with a virtually non-existent ground flora. The summer of 1976 proved to be the driest within living memory and drought conditions prevailed over much of England and Wales. These conditions brought about a fundamental change to the ecology of Lords Bushes.

A conspicuous feature of the soil profile is a relatively deep layer of leaf litter which in the normal course of events maintains a degree of moisture. This dried out in 1976 to such an extent that a fire would burn through the leaf litter, smouldering for days on end, occasionally bursting into flames and necessitating the attendance (often in shifts for days at a time) of the fire services, who did much good work in keeping the fires away from the surrounding houses. The dry conditions and smouldering leaf-litter affected the trees by creating a soil-moisture deficit and a number of trees lost their leaves. The smouldering leaf litter also burnt through the roots of the trees, in particular those of the beech, and thus caused the death of many trees. When dead beech trees, because of their shallow root system, (and compounded by the fact that old disused pollards tend to be top heavy) are very unstable in conditions of high wind. For safety reasons, because Lords Bushes is used as a public recreation area, the dead trees were felled using chain saws and heavy tracked vehicles, leaving a vast disturbed cleared area devoid of vegetation which I now refer to as 'the burnt and disturbed area' and it is the flora that has appeared on this site (Plate 2) that is discussed below. I censused the dead trees and stumps in January 1979 and found that the following trees had been killed and/or felled:

BEECH	OAK	HORNBEAM	BIRCH	HOLLY
165	52	33	36	484

These figures are probably an underestimate as many trees and stumps were burnt or removed. Of the felled trees, the following numbers were recorded as producing regrowth from the cut stool. In particular the remarkable regeneration powers of holly are of note.

BEECH	OAK	HORNBEAM	BIRCH	HOLLY
6	12	8	14	454

The flora that has appeared on the burnt and disturbed area can be divided into four main sections though there is of course overlap with some species:

WEED SPECIES. Obviously with any recently cleared area there is initially no competition for space, light or nutrients and such an area is readily exploited by what are traditionally considered to be weed species. The burnt and disturbed area of Lords Bushes is no exception and has been colonised by a great many weeds. The Compositae, including species such as creeping thistle (*Cirsium arvense*), coltsfoot (*Tussilago farfara*) and dandelion (*Taraxacum officinale*) all with wind blown seeds, are well represented on this area: likewise the willowherbs, in particular fireweed (*Epilobium angustifolium*), masters in the art of colonising waste ground. A number of species are ephemerals, including petty spurge (*Euphorbia peplus*) and groundsel (*Senecio vulgaris*), which have the ability to exploit rapidly an open waste area in its initial stages of colonisation, producing several generations in one year, but are unable to compete against the more vigorous species present. They disappear quickly from the area, as has happened in Lords Bushes, where I now find only the odd plant of groundsel growing, usually at the side of a path-way, and no petty spurge.

A number of species are of foreign origin. I was pleased to note the presence of five plants of gallant soldier (*Galinsoga parviflora*) in 1977, an uncommon weed originally from South America. Canadian fleabane (*Conyza canadensis*), a North American weed, and Oxford ragwort (*Senecio squalidus*), originally from volcanic slopes in the Mediterranean, region are two species that became common on the bomb sites of World War II London and are obviously well suited to the burnt and disturbed area of Lords Bushes.

GARDEN ESCAPES. The suburban areas which surround Lords Bushes have provided the source of a number of species, among them, cabbage (*Brassica oleracea*), flowering currant (*Ribes sanguineum*), gooseberry (*Ribes uva-crispa*), strawberry (*Fragaria x ananassa*) and columbine (*Aquilegia vulgaris*). Many of these will have been transported by birds but local people have planted species such as daffodil (*Narcissus* sp.) and scattered seeds of foxglove (*Digitalis purpurea*) on the burnt and disturbed area.

NATIVE TREE AND SHRUB SPECIES. This section includes only those that occur as seedlings and saplings and does not include the mature oak, hornbeam or other species that survived the drought and clearance.

Goat willow (*Salix caprea*) and birch (*Betula* spp.) are the major species of tree on this area, both species, as adults, produce many thousands of wind blown seeds and this probably accounts for their abundance on the burnt and disturbed area. Other native species occurring as seedlings include aspen, oak, beech and hornbeam usually in low numbers but oak is locally plentiful as a seedling, in some places the small seedlings carpeting the ground.

NATIVE HERBACEOUS PLANTS. This group of plants have provided some of the most interesting records. Undoubtedly a number of species in this group originated from the adjacent, undisturbed grassland. Among these are heath bedstraw (*Galium saxatile*) sheeps sorrel (*Rumex acetosella*) and heath grass (*Sieglingia decumbens*). (I have found only one tussock of the latter on the burnt and disturbed area and it is only a little more frequent in just two or the patches of undisturbed grassland). Others have undoubtedly originated outside Lords Bushes, among these are lesser stitchwort (*Stellaria graminea*), reed-mace (*Typha latifolia*) and bur-marigold (*Bidens tripartita*) the latter two species and a few others such as the buttercup (*Ranunculus flammula*), the uncommon (in Essex) water pennywort (*Hydrocotyle vulgaris*) and flote-grass (*Glyceria fluitans*) have all benefited from the generally wet conditions that have prevailed for much of the years 1977-1979. In contrast, the dry well-drained gravels exposed in places by the heavy plant during clearing, has provided habitat for an uncommon vetch, birdsfoot (*Ornithopus perpusillus*), which I found in just one confined area, though there were a number of plants.

Perhaps the most botanically interesting group of plants are *Juncus* (Rushes) and *Carex* (Sedges). Neither of these genera were well represented in Lords Bushes prior to 1976 yet now nine species of *Carex* are to be found on the burnt and disturbed area and six species of *Juncus* are to be found here. In the Flora of Essex (34) the only sedge to be recorded for Lords Bushes was black sedge (*Carex nigra*). Some of the sedges recorded are among the rarest to be found in Essex including star sedge (*Carex echinata*) and straight-beaked sedge (*Carex demissa*). I have recorded only a few specimens of each of these species on the burnt and disturbed area. Their appearance cannot satisfactorily be explained by bird transport or human introduction either deliberately or by chance. I feel that the only plausible explanation is that the seeds of these species have remained dormant in the soil, perhaps for many tens of years, and the gross disturbance during clearing may have prompted the germination of these seeds. This is almost certainly the case for some of the rushes, it is well established that species such as soft rush (*Juncus effusus*) lie dormant as seeds for many decades until the conditions for germination and subsequent growth are met with once again. The dormancy may also be applicable to species such as St Johns wort (*Hypericum pulchrum*) as this species was noted growing in Lords Bushes in the late nineteenth century (25).

It is noticeable that many of the species recorded from the burnt and disturbed area are typical of lowland heathland, among them the rushes *Juncus articulatus* and *J. squarrosus*, the sedges *Carex binervis*, *C. demissa* and *C. echinata*, the grass *Aira praecox*, birdsfoot vetch and water pennywort. If one takes into account the presence of ling (*Calluna vulgaris*) and other species such as heath bedstraw (*Galium saxatile*), heath grass (*Sieglingia decumbens*) and purple moor grass (*Molinia caerulea*) one has the representatives of what is probably a relict heathland flora. Many of these plants are likely to have occurred on the plains in Lords Bushes when actively managed as wood-pasture. The present flora compares very favourably with the flora of Tiptree Heath in Essex (27). Tiptree Heath is only a fraction of a former vast inter-manorial common but with far fewer trees than Epping Forest, heath being the dominant feature. It was, none the less, wood-pasture and this could explain the similarities in the flora.

(4) Flora of Knighton Lane

This area includes the narrow strip of disturbed ground that runs the length of Knighton Lane and also a small patch of grassland currently being invaded by oak scrub and bramble (possibly kept at bay until recently by rabbits, now not found in Lords Bushes). The ground bordering the road was I believe dug up during pipe-laying in the 1960's and the flora as a result of this is mainly a weed flora: Japanese knotweed (*Polygonum cuspidatum*) is increasing, while gorse (*Ulex europaeus*) has greatly declined and is now just about extinct here. Some holly was cleared from along this edge in 1980.

(5) Flora of the Northern Edge

There are large numbers of garden escapes and obviously deliberate plantings. This area abuts gardens along its entire length and rubble, garden refuse, and quite often whole plants, such as excess rhizomes of iris (*Iris germanica*) find their way into this edge of the woodland. Among the more interesting of these are summer snowflake (*Leucojum aestivum*) an uncommon introduction in Essex. Also uncommon in Essex, the sedge (*Cyperus longus*) is a fairly recent acquisition for Lords Bushes, apparently from a garden pond. Other typical garden plants found here include lilac (*Syringa vulgaris*), honesty (*Lunaria annua*), hollyhock (*Althea rosea*) and daisy (*Leucanthemum maximum*). Two typical garden weeds found along here are bindweed(*Calystegia sepium*) and another non-native, though long established, introduction is ground-elder (*Aegopodium podagraria*).

Some native plants are also of interest, lesser celandine (*Ranunculus ficaria*, occurring as both subspecies *bulbifer* and *ficaria*) is to be found growing in a great leafy mat to the east of the Kings Place entrance to Lords Bushes and also near here occurs the solitary primrose (*Primula vulgaris*) plant to be found in Lords Bushes. Ramsons (*Allium ursinum*) grows in a ditch either side of the entrance to a garden near the playground, and thus is possibly introduced. Perhaps the most enigmatic species in Lords Buhes is the rare (in Essex) wall lettuce (*Mycelis muralis*). Lords Bushes is apparently only the third known site in Essex for this species but there is a strong colony about the entrance to the old primary school.

(6) Flora of Forest Edge

This recording unit consists mainly of the trimmed roadside verge that runs the length of the Forest Edge site of Lord Bushes but includes the extreme eastern portion of the north edge. Underlying this area is London Clay and this almost certainly accounts for the presence of field maple (*Acer campestre*) and black bryony (*Tamus communis*), and probably also the sedge (*Carex divulsa*), which are more typical of basic soils. This edge of the wood, until about 1939, bordered fields and this may explain the presence of blackthorn (*Prunus spinosa*) and white campion (*Silence alba*) and there is also the possibility that the elm suckers (*Ulmus* sp.) originated from a tree planted along this edge.

Also of interest, was the presence in 1978 (not seen in 1979) of common cow-wheat (*Melampyrum pratense*) just inside the woodland near Squirrels Lane. This species became common for a short period when the undergrowth was cleared from this edge of Lords Bushes, following the building of houses along Forest Edge in the late 1930's (Scott pers. comm.)

(7) Flora of the Ponds

The ponds P1 and P2 are not a natural feature in Lords Bushes, they originated as gravel workings probably in the nineteenth century. Originally the workings were drained but on the cessation of gravel extraction these two pits were dammed forming the ponds we see today. From the photographs (Plates 3 and 4) taken from very nearly same position we can see how the flora has changed, due mainly to infilling by beech leaves and also the failure of the concrete dam. At one time the ponds must have had a healthy ecology (carp were caught in pond P2 in the 1940's (Scott pers. comm.) but now the flora and fauna are very poor reflecting the tendency to infill: present species include reed sweet grass (*Glyceria maxima*) sallow (*Salix cinerea*) and willowherb (*Epilobium hirsutum*). I have included in the lists two species found around the edge of the ponds: horsetail (*Equisteum arvense*) and gorse (*Ulex europaeus*), both these may have originated at the time the pits were being actively dug for gravel and were consequently more open.

CHECK-LIST OF THE VASCULAR PLANTS

The following check-list summarises the locations of the higher plants in Lords Bushes the numbers 1-7 are the recording units previously referred to and located on map 3(c).

Nomenclature follows that used in Jermyn, S.T. (34) with the exception of *Leycesteria formosa* Wall.

	1	2	3	4	5	6	7
EQUISETACEAE							
Equisetum arvense (Common Horsetail)							7
POLYPODIACEAE							
Dryopteris dilatata (Broad Buckler-fern)					5		
filix-mas (Male Fern)	1						
Pteridium aquilinum (Bracken)	1		3		5		
PINACEAE							
Pinus sp. (Pine)			3				
TAXACEAE							
Taxus baccata (Yew)	1		3				
RANUNCULACEAE							
Aquilegia vulgaris (Columbine)			3				
Ranunculus acris (Meadow Buttercup)			3				
bulbosus (Bulbous Buttercup)			3				
ficaria (Lesser Celandine)				4	5		
flammula (Lesser Spearwort)			3				
repens (Creeping Buttercup)			3	4	5		
sceleratus (Celery-leaved Crowfoot)			3				

PAPAVERACEAE							
Meconopsis cambrica (Welsh Poppy)					5		
Papaver somniferum (Opium Poppy)				4			
FUMARIACEAE							
Corydalis lutea (Yellow Fumitory)					5		
CRUCIFERAE							
Alliaria petiolata (Garlic Mustard)						6	
Brassica oleracea (Cabbage)			3				
Capsella bursa-pastoris (Shepherds Purse)			3	4			
Cardamine flexuosa (Wavy Bitter-Cress)			3				
hirsuta (Hairy Bitter-Cress)			3				
Coronopus didymus (Lesser Swine-cress)			3				
Lunaria annua (Honesty)					5		
Sinapis arvensis (Charlock)				4			
Sysymbrium officinale (Hedge Mustard)				4			
VIOLACEAE							
Viola riviniana (Dog Violet)					5	6	
tricolor (Heartsease)			3				
HYPERICACEAE							
Hypericum humifusum (Trailing St John's Wort)			3				
pulchrum (Slender St John's Wort)			3				
CARYOPHYLLACEAE							
Cerastium holosteoides (Mouse-ear Chickweed)			3				
Moerhingia trinervia (Three-nerved Sandwort)	1		3				
Sagina procumbens (Procumbent Pearlwort)					5		
Silene alba (White Campion)			3			6	
dioica (Red Campion)			3			6	
Stellaria alsine (Bog Stitchwort			3				
graminea (Lesser Stitchwort)			3				
holostea (Greater Stitchwort)						6	
media (Common Chickweed)			3	4			
CHENOPODIACEAE							
Chenopodium album (Fat Hen)			3				
polyspermum (Many-seeded Goosefoot)			3	4			
MALVACEAE							
Althaea rosea (Hollyhock)					5		
Malva sylvestris (Common Mallow)				4			
GERANIACEAE							
Geranium robertianum (Herb Robert)					5		
OXALIDACEAE							
Oxalis corymbosa (Oxalis)					5		
europaea (Oxalis)					5		
ACERACEAE							
Acer campestre (Field Maple)		2					
pseudoplatanus (Sycamore)		2	3	4			
HIPPOCASTANACEAE							
Aesculus hippocastanum (Horse Chestnut)					5		
AQUIFOLIACEAE							
Ilex aquifolium (Holly)		2	3				
VITACEAE							
Parthenocissus sp. (Virginia Creeper)					5		
LEGUMINOSAE							
Laburnum anagyroides (Laburnum)					5		
Lotus pedunculatus (Birdsfoot Trefoil)			3				
Ornithopus perpusillus (Birdsfoot)			3				
Sarothamnus scoparius (Broom)						6	
Trifolium campestre (Hop Trefoil)			3				
repens (White Clover)			3	4			
Ulex europaeus (Gorse)			3				7
Vicia hirsuta (Hairy Tare)			3				
sativa (Common Vetch)			3			6	
ROSACEAE							
Amelanchier sp. (Mespilus)	1						
Cotoneaster sp. (Cotoneaster)			3				
Crataegus laevigata (Midland Hawthorn)	1						

monogyna (Common Hawthorn)	1						
Fragaria x ananassa (Garden Strawberry)			3				
vesca (Wild Strawberry)			3				
Geum urbanum (Herb Bennet)				4			
Malus sylvestris spp. sylvestris (Crab Apple)	1		3				
Potentilla erecta (Common Tormentil)		2	3				
sterilis (Barren Strawberry)						6	
Prunus avium (Cherry)						6	
domestica agg. (Plum)					5		
laurocerasus (Cherry Laurel)	1		3		5		
spinosa (Blackthorn)						6	
Rosa sp. (Rose)	1		3			6	
Rubus fruticosus agg. (Bramble)	1		3	4	5		
idaeus (Raspberry)			3		5		
Sorbus aucuparia (Rowan)	1		3				
intermedia (Swedish Whitebeam)	1		3				
torminalis (Wild Service)	1						
Spiraea sp. (Spiraea)					5		
GROSSULARIACEAE							
Ribes sanguineum (Flowering Currant)			3				
uva-crispa (Gooseberry)			3				
ONAGRACEAE							
Circaea lutetiana (Enchanters Nightshade)	1				5		
Epilobium adenocaulon (American Willow-herb)			3		5		
angustifolium (Fireweed)			3		5		
hirsutum (Great Hairy Willow-herb)			3		5		7
montanum (Broad-leaved Willow-herb)			3				
roseum (Small-flowered Willow-herb)			3				
CALLITRICHACEAE							
Callitriche stagnalis (Common Water Starwort)			3				
ARALIACEAE							
Hedera helix (Ivy)	1		3	4		6	
UMBELLIFERAE							
Aegopodium podagraria (Goutweed)					5		
Aethusa cynapium (Fools Parsley)				4			
Anthriscus sylvestris (Cow Parsley)				4		6	
Heracleum sphondylium (Hogweed)			3	4			
Hydrocotyle vulgaris (Pennywort)			3				
Pastinaca sativa (Wild Parsnip)						6	
Torilis japonica (Upright Hedge Parsley)						6	
EUPHORBIACEAE							
Euphorbia lathyrus (Caper Spurge)					5		
peplus (Petty Spurge)			3		5		
Mercurialis annua (Annual Mercury)					5		
POLYGONACEAE							
Polygonum aviculare (Knotgrass)			3	4	5		
convolvulus (Black Bindweed)			3				
cuspidatum (Japanese Knotweed)				4	5		
persicaria (Redshank)			3				
Rumex acetosa (Sorrel)						6	
acetosella (Sheep's Sorrel)		2	3				
obtusifolius (Broad-leaved Dock)			3	4	5		
sanguineus (Red-veined Dock)			3				
URTICACEAE							
Urtica dioica (Stinging Nettle)			3	4	5		
urens (Annual Nettle)			3				
CANNABIACEAE							
Humulus lupulus (Hop)					5		
ULMACEAE							
Ulmus sp. (Elm)						6	
BETULACEAE							
Betula pendula (Silver Birch)	1		3				
pubescens (Birch)	1		3				
CORYLACEAE							
Carpinus betulus (Hornbeam)	1		3				

FAGACEAE							
Fagus sylvatica (Beech)	1		3				
Quercus petraea (Sessile Oak)	1						
robur (Common Oak)	1		3				
SALICACEAE							
Populus tremula (Aspen)	1		3				
Salix caprea (Goat Willow)			3				
cinerea agg. (Common Sallow)			3				
ERICACEAE							
Calluna vulgaris (Ling)			3				
PRIMULACEAE							
Anagallis arvensis (Scarlet Pimpernel)					5		
Primula vulgaris (Primrose)					5		
BUDDLEJACEAE							
Buddleja davidii (Buddleia)			3				
OLEACEAE							
Fraxinus excelsior (Ash)	1						
Syringa vulgaris (Lilac)					5		
BORAGINACEAE							
Myosotis sylvatica (Wood Forget-me-not)					5		
Pentaglottis sempervirens (Alkanet)				4			
Pulmonaria officinalis (Lungwort)					5		
CONVOLVULACEAE							
Calystegia sepium (Great Bindweed)					5		
SOLANACEAE							
Solanum dulcamara (Bittersweet)			3		5		7
nigrum (Black Nightshade)			3				
tuberosum (Potato)			3				
SCROPHULARIACEAE							
Antirrhinum majus (Snap Dragon)			3		5		
Cymbalaria muralis (Ivy-leaved Toadflax)					5		
Digitalis purpurea (Foxglove)	1		3	4	5		
Linaria purpurea (Purple Toadflax)					5		
Melampyrum pratense (Common Cow-wheat)						6	
Scrophularia vernalis (Yellow Figwort)					5		
nodosa (Figwort)	1						
Veronica arvensis (Wall Speedwell)			3				
· *hederifolia* (Ivy-leaved Speedwell)				4	5		
LABIATAE							
Ballota nigra (Black Horehound)			3	4	5		
Glechoma hederacea (Ground Ivy)				4		6	
Lamium album (White Dead-nettle)				4			
maculatum (Spotted Dead-nettle)			3				
purpureum (Purple Dead-nettle)			3	4			
Lycopus europaeus (Gypsy-wort)			3				
Mentha sp. (Mint)					5		
Prunella vulgaris (Self-heal)					5		
Stachys sylvatica (Hedge Woundwort)			3				
PLANTAGINACEAE							
Plantago lanceolata (Ribwort Plantain)						6	
major (Great Plantain)			3	4	5		
RUBIACEAE							
Galium aparine (Goosegrass)			3	4	5		
saxatile (Heath Bedstraw)		2	3				
CAPRIFOLIACEAE							
Leycesteria formosa			3				
Lonicera peryclymenum (Honeysuckle)	1					6	
Sambucus nigra (Elder)	1		3	4		6	
COMPOSITAE							
Achillea millefolium (Yarrow)			3			6	
Arctium minus (Lesser Burdock)				4		6	
Artemesia vulgaris (Mug wort)			3	4	5		
Aster sp. (Michaelmas Daisy)			3				
Bidens tripartita (Bur-marigold)			3				
Bellis perennis (Daisy)	1						

Carduus acanthoides (Welted Thistle)			3				
Cirsium arvense (Creeping Thistle)			3				
vulgare (Spear Thistle)			3				
Conyza canadensis (Canadian Fleabane)			3				
Crepis capillaris (Smooth Hawk's-beard)			3				
Galinsoga parviflora (Gallant Soldier)			3				
Helianthus annuus (Sunflower)			3				
Hypochoeris radicata (Cat's Ear)		2	3				
Lactuca serriola (Prickly Lettuce)			3				
Lapsana communis (Nipplewort)			3	4			
Leontodon autumnalis (Autumnal Hawkbit)		2				6	
Leucanthemum maximum (Shasta Daisy)					5		
vulgare (Marguerite)				4	5		
Matricaria matricarioides (Mayweed)			3	4			
Mycelis muralis (Wall Lettuce)					5		
Senecio erucifolius (Hoary Ragwort)						6	
jacobaea (Common Ragwort)			3				
squalidus (Oxford Ragwort)			3				
vulgaris (Groundsel)			3	4			
Solidago canadensis (Golden Rod)			3				
Sonchus asper (Spiny Sow-thistle)			3				
oleraceus (Sow-thistle)			3				
Tanacetum parthenium (Feverfew)					5		
Taraxacum officinale sens. lat. (Dandelion)			3	4			
Tragopogon pratense (Goats-beard)						6	
Tripleurospermum maritimum (Scentless Mayweed)			3				
Tussilago farfara (Colts-foot)			3				
LILIACEAE							
Allium ursinum (Ramsons)					5		
Endymion non-scriptus (Bluebell)	1			4	5		
Ruscus aculeatus (Butchers Broom)	1						
JUNCACEAE							
Juncus articulatus (Jointed Rush)			3				
bufonius (Toad Rush)	1		3				
bulbosus (Bulbous Rush)			3				
effusus (Soft Rush)	1		3				
squarrosus (Heath Rush)			3				
subuliflorus (Common Rush)			3				
Luzula campestris (Field Woodrush)		2					
multiflora (inc. F. *conglomerata*) (Heath Woodrush)			3				
pilosa (Hairy Woodrush)	1						
AMARYLLIDACEAE							
Leucojum aestivum (Summer snowflake)					5		
Narcissus sp. (Daffodil)			3			6	
IRIDACEAE							
Crocus sp. (Crocus)					5		
Iris germanica (Flag Iris)					5		
pseudacorus (Yellow Flag)							7
DIOSCORIACEAE							
Tamus communis (Black Bryony)						6	
ARACEAE							
Arum maculatum (Cuckoo-pint)					5		
LEMNACEAE							
Lemna minor (Duckweed)							7
TYPHACEAE							
Typha latifolia (Great Reedmace)			3				
CYPERACEAE							
Carex binervis (Ribbed Sedge)			3				
demissa (Straight-beaked Sedge)			3				
divulsa (Grey Sedge)						6	
echinata (Star Sedge)			3				
nigra (Black Sedge)	1		3				
ovalis (Oval Sedge)			3				
pendula (Pendulous Sedge)			3				

pilulifera (Pill Sedge)			3				
remota (Remote Sedge)	1		3				
sylvatica (Wood Sedge)			3				
Cyperus longus (Galingale)					5		
GRAMINEAE							
Agropyron repens (Couch-grass)					5		
Agrostis stolonifera (Creeping Bent-grass)		2	3		5		
tenuis (Common Bent-grass)		2	3		5		
Aira praecox (Early Hair-grass)			3				
Alopecurus pratensis (Meadow Foxtail)			3				
Arrhenatherum elatius (False Oat-grass)			3			6	
Brachypodium sylvaticum (Slender False-brome)					5		
Bromus mollis (Soft Brome)			3				
sterilis (Barren Brome)			3	4	5		
Cynosurus cristatus (Crested Dogs-tail)			3				
Dactylis glomerata (Cocks-foot)		2	3	4	5		
Deschampsia caespitosa (Tufted Hair-grass)			3				
flexuosa (Wavy Hair-grass)		2					
Festuca rubra (Red Fescue)		2	3				
Glyceria fluitans (Flote-grass)			3				7
maxima (Reed-grass)							7
Holcus lanatus (Yorkshire Fog)		2	3		5		
mollis (Creeping Soft-grass)		2	3		5		
Hordeum murinum (Wall Barley)			3	4	5		
Lolium perenne (Rye-grass)		2	3		5		
Molinia caerulea (Purple Moor-grass)	1		3				
Phleum pratense (Timothy-grass)			3		5		
Poa annua (Annual Meadow Grass)	1	2	3	4	5		
nemoralis (Wood Poa)	1		3				
pratensis (Meadow Grass)	1	2	3		5		
Sieglingia decumbens (Heath-grass)		2	3				

NOTES ON TREE AND SHRUB SPECIES

Lords Bushes contains at least 40 species of tree and shrub. This list is undoubtedly enhanced by the presence of a high proportion of non-native species which have originated in nearby gardens or may possibly have been transported as seed some distance for example by birds. In particular, the northern edge of Lords Bushes has a number of introduced tree and shrub species, many of these including plum (*Prunus domestica*), cherry (*Prunus avium*) and lilac (*Syringa vulgaris*) appear to have originated as suckers from the parent tree or bush in the adjacent garden.

The burnt and disturbed area has provided some notable records, in particular the abundance of small trees of goat willow (*Salix caprea*), a species noted often as an early coloniser of waste sites, and similarly a number of buddleia (*Buddleja davidii*) have appeared on the area.

Sycamore (*Acer pseudoplatanus*) occurs only in a relatively small area on the Knighton Land side of the undisturbed woodland (bar a solitary seedling found on the burnt and disturbed area) on ground possibly cleared originally in the nineteenth century, following gravel extraction. Mountain Ash (*Sorbus aucuparia*) is not common as a mature tree but well distributed mainly on the margin of Lords Bushes and in the old gravel workings. It is traditionally considered to be an introduction in Essex, though many trees undoubtedly arise from bird sown seeds, as no doubt have the seedlings of this species found on the burnt and disturbed area.

Swedish whitebeam (*Sorbus intermedia*) is to be found only as a single mature standard N.E. of pond P2; this tree has undoubtedly given rise to the seedlings found on the burnt and disturbed area. Ash (*Fraxinus excelsior*) likewise is to be found only as a single mature tree on the Knighton Lane edge of Lords Bushes it is a tree incompatible with wood-pasture and probably was able to establish itself only on the cessation of grazing in Lords Bushes, interestingly it shares its root-base with a hornbeam tree.

Yew (*Taxus baccata*) occurs as a single small tree in the undisturbed woodland near Squirrels Lane and also on the burnt and disturbed area as seedlings. It is a rare tree in Epping Forest, probably because it is poisonous to livestock and efforts would have been made to abolish it from the Forest.

The Native Tree and Shrub Species

Lords Bushes contains a number of interesting native tree and shrub species. Wild service would seem to underline the long established nature of the woodland.

Beech *(Fagus sylvatica)*

Prior to the drought of 1976, this was undoubtedly the dominant tree species in Lords Bushes. The tall beech trees, until 1976, were overshading and thus killing other tree species particularly the oak pollards; the beech tending to become the sole dominant in the woodland. In the undisturbed woodland beech occurs mainly as a standard and pollard but there are several trees that have been coppiced one of these being a particularly fine example, (Plate 11.) which shows evidence that an attempt was made to pollard the coppice poles. Perhaps this is evidence of an unsuccessful attempt at introducing coppicing (coppicing is easier than pollarding) into a wood-pasture system.

Also of note in Lords Bushes is the presence of the beech scale insect (*Cryptococcus fagi*), a true bug (Hemiptera: Coccidae) specific to beech. The insects presence is characterised by a 'woolly' protective wax scale under which the colonial females live. They are parthenogenetic, no males are known. Heavy infestations of the scale insect often occur on overmature trees, especially those weakened in any way. The infestation can be severely debilitating, but it is this insect's association with a pathogenic fungus of the genus *Nectria* which, in many parts of Britain, notably in the Chilterns, has led to the deaths of many trees. Infestations of the insect are not invariably followed by fungal infection (47).

Oak *(Quercus robur)*

All oaks bar one so far examined are English oak (*Quercus robur*). The exception appears to have predominantly sessile oak (*Quercus petraea*) characteristics, namely the acorn cups are sessile, in most cases flush against the stem, while the leaves have relatively long petioles and lack well developed auricles (in many leaves this feature is totally lacking). It is interesting to note that Warner writing in 1771 (74) records the "Oak tree with the acorns on the short shoot stalks found on the Forest between Muncombe and the Bald Faced Stag" and it is thus likely that more diligent searching would turn up further examples of this species. Mature oaks occur in Lords Bushes only as standards and pollards, though the latter are declining due to overshading by beech. Oak is a light demanding species (44) and presumably, in a pollard woodland where the majority of trees in a given area are pollarded at the same time, there is no competition for light between individual trees for much of the time and hence the oak trees are able to survive. E. N. Buxton wrote of the oaks in Lords Bushes "It is the old oak pollards which give it a character of its own . . . many of these have been killed by the overshadowing of the lusty young beeches but others have been taken in time and saved from the slow but sure process of choking . . . no one can doubt that the ancient boles will continue to give a venerable air to this wood for many years to come" (14).

Perhaps the finest of these trees is the Pulpit Oak (Plate 5) situated in the far north west corner of the wood, bordering Monkhams Lane. This magnificent tree, which has a girth at breast height of 4.57m (15ft 9½"), could well be over 600 years old and is by its sheer size set apart from the other oaks in Lords Bushes. I can find no written evidence as to the origin of the name 'Pulpit Oak' but it seems likely that the tree could be what is popularly known as a 'Gospel Oak', where the gospel was read at the Rogation-tide perambulation and the parish bounds beaten (31). Although not precisely on the boundary, the Pulpit Oak in bordering Monkhams Lane is situated near what was anciently the boundary of the parish of Chigwell, which was probably of Saxon origin.

Interestingly, Monkhams Lane is still an important boundary: that of Epping Forest District Council and it is also concurrent with part of the modern Essex County boundary. Another oak tree known in the 19th century as the 'Poachers Pocket' is also recorded for Lords Bushes but I have been unable to find this tree. (62).

Hornbeam *(Carpinus betulus)*

This tree much favoured by the commoners for pollarding is indeed to be found commonly as a pollard in Lords Bushes, though because of the characteristic contorted appearance it assumes on pollarding, it was the target of the early Conservators who considered them not to be very aesthetically pleasing (6) and consequently many were felled (Plate 8). A number were felled after the drought of 1976, though presumably because it is rather deeper-rooted it did not suffer to the same extent as the beech.

Aspen *(Populus tremula)*

Aspen is uncommon as a mature tree in Lords Bushes, occurring at just two points on the edge of one of the footpaths in the undisturbed woodland. I have found a number of saplings on the burnt and disturbed area, testament to the generally wet years after 1976. Its foliage is not liked by livestock, hence its survival in wood-pasture (59).

Holly *(Ilex aquifolium)*

Holly, in the undisturbed woodland of Lords Bushes as in much of Epping Forest, dominates the understorey tolerating the deep shade cast by beech trees. This feature seems to be characteristic of derelict wood-pasture and is probably caused by lack of browsing (58).

Midland Hawthorn *(Crataegus laevigata)*

This shrub is generally considered to be a species indicating a long established woodland (59). It is not common, but it does occur in the vicinity of the wild service in the S.E. corner of Lords Bushes. This species has a distinctive leaf shape and the presence of two seeds in the berry identifies midland hawthorn though there appears also to be much hybridisation with the common hawthorn.

Field Maple *(Acer campestre)*

Maple occurs mainly as a standard in Lords Bushes, though one tree has been coppiced. Interestingly this species is confined to the London Clay in Lords Bushes (see map 4(a)).

Crab Apple *(Malus sylvestris)*

Again occurring mainly as a standard (there is one tree on the Forest Edge side of Lords Bushes which has been coppiced). All specimens of this tree are the native glabrous sub-species *sylvestris*, most of the trees are rather tall and spindly; probably it is another species that is being forced to compete with the excessive overshading of the beech trees (see map 4 (b)).

Ivy *(Hedera helix)*

Although in reality a woody climber, I have treated this species as a 'shrub'; it is generally infrequent in Lords Bushes though along the Knighton Lane edge of the undisturbed woodland, beneath a stand of sycamore, and by the pond P2 this species

Plate 1. Holly and tree clearance on the north side of Lords Bushes (1979). M. A. Hanson.

Plate 2. View over part of the burnt and disturbed area (1978). S. A. Quyyumi.

Plate 3. Pond P2. Note the dense growth of reed-grass (*Glyceria maxima*) contrast this with Plate 4 (1979). M. A. Hanson.

Plate 4. Pond P2. Taken c. 1913.

Plate 5. The Pulpit Oak (1979). R. Green.

Plate 6. Aerial photographs of Lords Bushes taken in April (right) and May (left) 1976 shortly before the draught and subsequent clearance. Note density of tree cover. By permission of Epping Forest District Council.

Plate 7. Root system of overturned Beech (1979). M. A. Hanson.

Plate 8. A hornbeam pollard. One of the earliest photographs of Lords Bushes taken in March 1894 by Hugh Main.

Plate 9. Squirrels Lane (tape shows depression). Large tree trunk, right background is Wild Service No. 2. (1979). M. A. Hanson.

Plate 10. The experimental tramway in Lords Bushes 1872/3.

Plate 11. Coppiced Beech. Note the coppice poles have been pollarded (1979). M. A. Hanson.

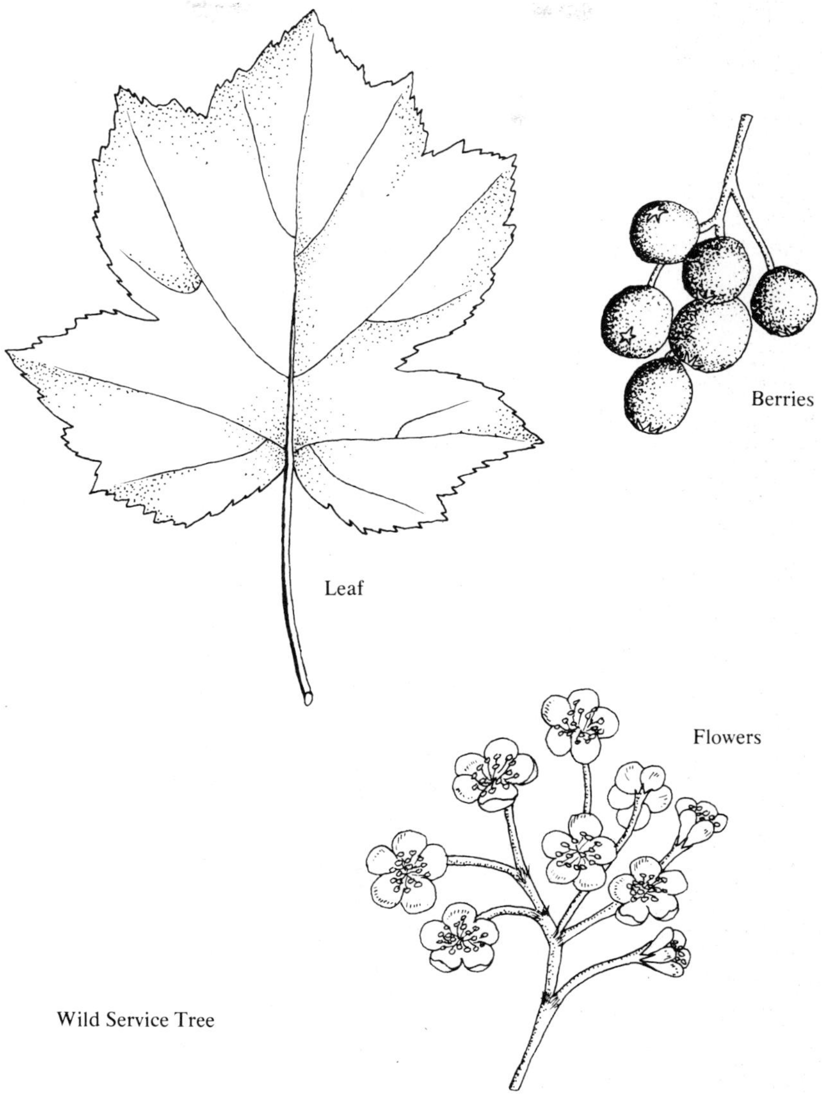

Wild Service Tree

Map 4. Distribution of some tree species in Lords Bushes.

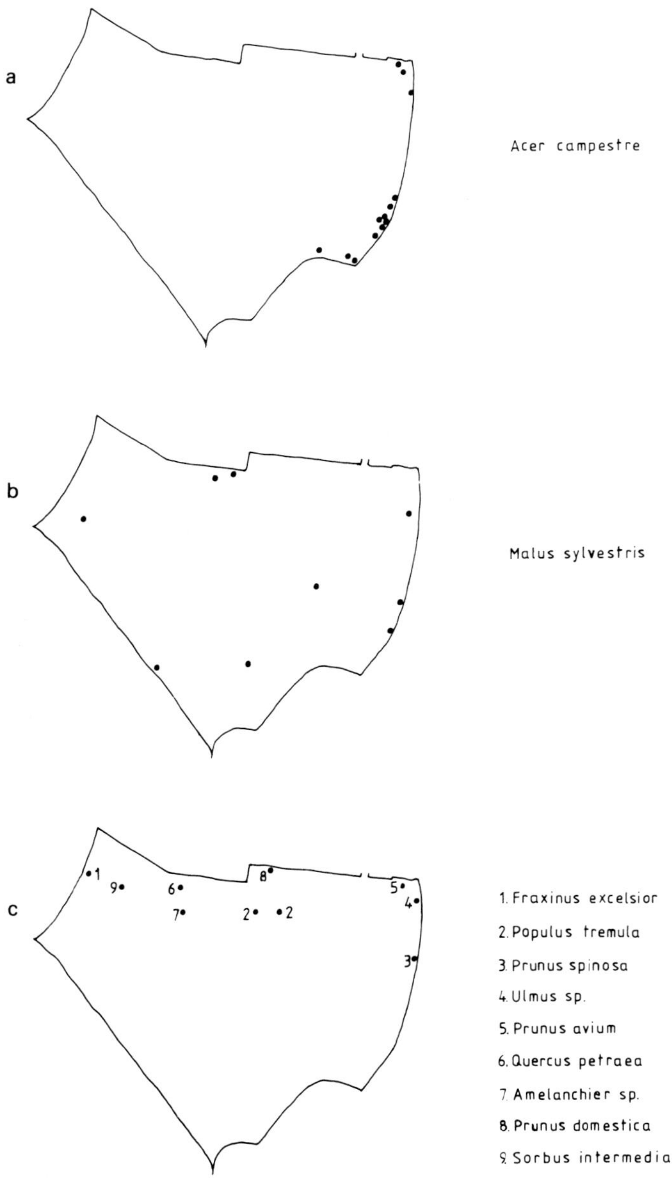

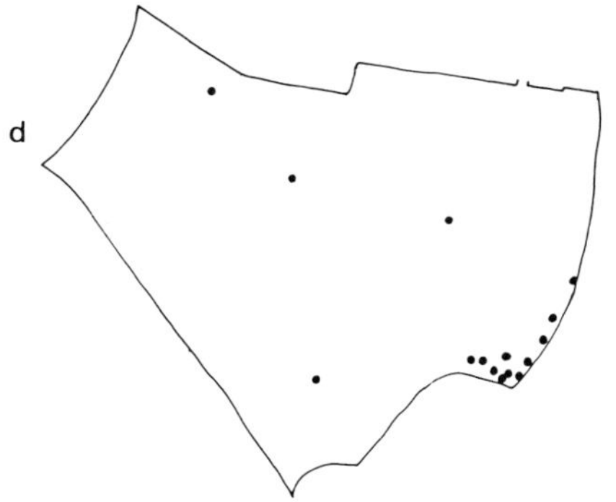

Sorbus torminalis

carpets the ground. This is a feature typical of secondary woodland (58), though paradoxically it also occurs in two or three other areas carpeting the ground in the undisturbed woodland.

Wild Service *(Sorbus torminalis)*

This tree is undeniably the most outstanding feature of Lords Bushes. It is a rare tree whose presence is taken to indicate an ancient continuity of woodland. In Britain wild service has largely a southern distribution and in eastern England, Essex has by far the greatest concentration of this tree (59). Most records come from single trees in woodland but there are also a number of records from hedgerows on Tiptree Heath and in the Rayne, Black Notley and Felsted area (34). Epping Forest has a relatively large population of this species and Lords Bushes is probably the finest site for this species in the Forest.

The overall distribution in Lords Bushes indicates a concentration on the London Clay in the extreme south east corner with just four trees isolated from the main group, three of which are growing on Pleistocene sands and gravels while the third is growing the side of drainage ditch on exposed London Clay. Notably many of the trees grow on or near a pathway or near a brook or ditch. Lloyd (42) in his paper on service in Epping Forest also noted this. In Monks Wood Huntingdonshire service tends to grow on the margin or close by the borders of rides (63) and along ditches, rarely does it appear in the middle of a stand of trees, possibly semi-shade is an ecological requirement of this species.

What is most unusual about the distribution of one group of wild service in Lords Bushes is that the trees closely follow the banks of the old Squirrels Lane. The presence of these trees may indicate the antiquity of this trackway (Plate 9). I can find no reference to the origin of the name 'Squirrel' (if indeed this was the original name) though two fields, which are now built on but were located immediately south of the section of Squirrels Lane in Lords Bushes, are named in the tithe award of 1838 as 'Further Squirrels' and 'Hither Squirrels' (68).

It is unlikely the trees present today were planted and the trees have in all probability just suckered along the banks of the road. Lloyd (42) also records wild service on the boundary bank of the old drove road, Blind Lane, near Lippitts Hill and Richard Warner (74) records wild service thus, "In woods and hedges, found in Sakes commonly called Snakes Lane. On a bank between Mr Warners Fields and Ribton Lane [probably a boundary bank] and in a wood called the Sale in the Parish of Walthamstow.

Further work on leaf configuration (assuming the trees have suckered and hence the suckers having similar leaves to the parent tree) could shed some light on the parentage and distribution of wild service in Lords Bushes. At least two of the mature trees have borne chequers (the berry of wild service, also known colloquially in Essex as 'sarves'), much to the delight of a flock of blackbirds.

Whatever the origin of these trees, Lords Bushes and indeed Epping Forest is fortunate to possess them and it is also fortunate that when the land immediately east of Lords Bushes was cleared for agriculture the clearance did not penetrate any further into Lords Bushes otherwise this could have removed the last vestige of this fine tree from the woodland. It is interesting to speculate that wild service may have been a frequent tree on the lower slopes of the Roding Valley before its clearance for cultivation.

LOWER PLANTS

Mosses

Essex, lacking in rock exposures or a variety of relief and being one of the least humid counties in Great Britain, is not a 'good' county for Bryophytes. In Lords Bushes, the close proximity of urban and industrial areas with their associated pollution precludes the presence of many species of bryophyte. I have records of just 22 species of moss (I have recorded only on the burnt and disturbed area of Lords Bushes) noted mostly in the winter of 1978–79.

The burnt and disturbed area with its open ground, exposed gravels, numerous tree stumps and fallen trees has provided a great variety of micro-habitat for colonisation by mosses. Of particular interest are the *Polytrichum* spp, which have undoubtedly benefited from the clearance (they do not appear to be present in the undisturbed woodland but are of very frequent occurrence on the burnt and disturbed area). Apart from in Epping Forest this genus, typical of acid substrates, is uncommon in Essex. The acid nature of Lords Bushes is further underlined by the presence of *Tetraphis pellucida* and indeed the majority of mosses recorded are typically found in woodlands of low pH. Another micro-habitat are the bonfire sites which were readily colonised by great mats of the moss *Funaria hygrometrica* though this species also occurred extensively on the exposed gravels.

The species listed below and the micro-habitat is indicated by a letter thus:

G Ground
B Bark (tree species in brackets)
D Decaying wood
F Bonfire site

Nomenclature for Mosses and Liverworts follows Jermyn S. T. (34).

Atrichum undulatum	G	
Aulacomnium androgynum	D	
Brachythecium rutabulum	G	
velutinum	G	
Campylopus introflexus	G	
Ceratodon purpureus	G	
Dicranella heteromalla	G	
Dicranoweisia cirrata	B	(Beech)
Dicranum scoparium	B	(Beech stump)
Eurhynchium praelongum	D,B	(Hornbeam)
Funaria hygrometrica	F,G	
Hypnum cupressiforme	G,B	(Hornbeam)
Isopterigium eligans	G	
Mnium hornum	G	
Orthodontium lineare	D	
Plagiothecium curvifolium	B	(Hornbeam)
Pohlia nutans	G	
Polytrichum commune	G	
formosum	G	
juniperinum	G	
piliferum	G	
Tetraphis pellucida	B	(Birch stump)

Liverworts

The variety of liverworts is limited by the same factors that produce a restricted moss flora. So far I have records of just six species. *Marchantia polymorpha* was abundant on the burnt and disturbed area in 1977, forming mats of thalli of up to one square metre. Its abundance was short-lived and by 1979 it was rather uncommon. *Lophocolea heterophylla* is said to be the commonest ephiphyte in Essex (34), occurring well into polluted areas.

Calypogeia fissa
Cephalozia bicuspidata
Gymnocolea inflata
Lophocolea heterophylla
Marchantia polymorpha
Pellia epiphylla

Lichens

It is well established that lichens are susceptible to pollution, in particular to sulphur compounds. Lords Bushes is surrounded by urban development and with industrial areas in close proximity, so it is no surprise that I have recorded just five species of lichen (though it is likely further, less conspicuous, species will be recorded).

Because of the pollution, Lords Bushes lacks any of the species associated with ancient woodland such as *Parmelia revoluta*. On the Pulpit Oak (probably the oldest tree in Lords Bushes) I have been able to record only *Lepraria incana* on the trunk; in a low pollution area this tree would undoubtedly be covered in lichens of many species.

Excluding saxicolous species, the following were recorded for Lords Bushes in 1979. Nomenclature follows Hawksworth, D. L., James, P. W. & Coppins, B. J. (32).

Cladonia coniocraea Fairly common on the lower trunks of trees including beech and silver birch, also on decaying logs.
Cladonia rangiformis Found only on the ground by pond P2.
Hypogymnia physodes Uncommon, only a single immature thallus noted on Silver Birch.
Lecanora conizaeoides Common, noted on oak.
Lepraria incana Common, noted on many trees of a variety of species.

Fungi (by G. Kibby)

Lords Bushes as a whole is quite different in character from most of Epping Forest and a lot smaller, as well as more limited in the variety of habitats. As a consequence the mycological flora is a lot more restricted in terms of species, although in a good year the wood can be very productive of fruit bodies.

The recent major fires which swept the woodland during and after the drought of 1976, plus the subsequent removal of a large number of the trees, has presented an unparalleled opportunity to study the advent of carbonicolous species and the change in overall flora as the woodland regenerates. Within the first year following the fire, vast numbers of typical burnt area species began making their appearance. Most notable were the thousands of fruit-bodies of *Pholiota highlandensis* (=*carbonaria* pp.) included among which were several clumps of what appear to be a closely related but clearly different species. A confident determination was not possible at the time although it seems likely to have been the true *P. carbonaria* known from a few other sites in Britain. Further collections are needed to be certain.

Daldinia concentrica, a hard black pyrenomycete, also appeared in large numbers and this seems to specialise in scorched birch trees, pushing its way through cracks in the bark.

Not included on the list, and indeed not seen again since a large number of fruit bodies appeared in 1972, is a bolete close to *erythropus* but with a pitch black cap and deepest red pores which is as yet undescribed although it has been illustrated by Pilat (56). It was growing on the central path which divides Lords bushes and Knighton woods

Boletus erythropus

and rather unusually for a bolete on very wet, gravelly soil. It is hoped that it is still growing in the area and perhaps waiting for a particularly good season to fruit again.

Some of the very damp parts of the wood will repay closer examination and I feel many interesting species still await discovery. Most notable are the often large numbers of *Boletus parasiticus* to be found parasitising the common earth-ball, *Scleroderma citrinum*. This is a distinctly uncommon to rare species almost entirely restricted to the south of England, we appear to be particularly fortunate in that it is rather common in Epping Forest. *Inocybe fastigiata* appears to be commoner in these wet hollows than in Epping Forest, where only a few records have been made. this is a genus where more records are likely as they are common in the area and modern keys are now available for determining them (45).

Hebelomas are abundant in this woodland in late autumn, as they are in most woodlands in Britain, but they must represent one of the most baffling and unsatisfactory genera as far as determinations are concerned. They present a minimum of useful taxonomic characters and all look very similar indeed; however, more complete keys are appearing in European mycological journals and it is hoped that these will enable at least some of them to be named. We have four species on the list and I estimate at least another four are present in the wood, almost certainly the carbonicolous *H.carbonicola* will make an appearance sometime soon with the abundance of now suitable habitat. It is recognisable by its tough, very elastic flesh and reddish brown colours.

Russulas although numerous are mostly common species except for *R.romelli* which has been collected here but not elsewhere in Epping Forest.

I would estimate that the list could be increased by at least fifty per cent by means of diligent collecting and research using up-to-date literature and collecting at *all times of the year*, something many mycologists sadly do not do. Although not a site of great mycological importance the wood is nevertheless interesting and because of its relatively small size allows the possibility of a really thorough study over a few years.

Nomenclature follows Corner, E. J. H. (16); Dennis, R. W. G., Orton, P. D. and Hora, F. B. (19); Pegler, D. N. (53) and Watling, R. (75).

BOLETALES

Boletus parasiticus
B.erythropus
B.versicolor
B.appendiculatus
B.porosporus
B.badius
B.chrysenteron
B.subtomentosus
B.pruinatus
Leccinum scabrum
L.variicolor
L.versipelle
Paxillus involutus

AGARICALES

Pholiota highlandensis
P.squarrosa
P.ochrochlora
Inocybe fastigiata
I.lanuginella
I.napipes
Cortinarius decipiens
C.anomalus
Dermocybe semisanguineus
Hebeloma sacchariolens
H.crustuliniforme
H.longicaudum
H.mesophaeum
H.testaceum
Aminata phalloides
A.rubescens
A.citrina
A.spissa
A.fulva
A.muscaria
Lepiota rhacodes
L.cristata
Agaricus arvensis
Armillaria mellea
A.polymyces
A.tabescens
Coprinus lagopoides
C.atrementarius
C.micacius
Mycena galericulata
M.inclinata
M.leucogala
M.polygramma
M.vitilis
Entoloma rhodopolium
Nolanea sericeum
Gymnopilus penetrans
G.junonius
Pleurotus ostreatus
P.pulmonarius
P.dryinus
P.salignus
Panellus stipticus
Hypholoma fasciculare
Lyophyllum decastes
Tricholoma saponaceum
Clitocybe vibecina
C.inversa
C.infundibuliformis
Lepista nudum
Collybia fusipes
C.peronata
C.dryophila
Laccaria laccata
L.proxima
L.amethystea
Pluteus cervinus
P.salicinus
Flammulina velutipes
Russula vesca
R.xerampelina
R.romelli
R.mairei
R.fragilis
R.rosea
R.cyanoxantha
R.ionochlora
R.puellaris
R.atropurpurea
R.aeruginea
R.knauthii
Lactarius subdulcis
L.tabidus
L.quietus
L.rufus
L.vietus
L.glyciosmus
Tricholomopsis platyphylla
Melanoleuca melaleuca
Schizophyllum commune
Crepidotus variablis
Tubaria furfuracea
Panaeolina foenisecii

APHYLLOPHORALES

Clavulina cristata
C.cinerea
Thelephora terrestris
Stereum hirsutum
Cantherellus cibarius
Fistulina hepatica
Piptoporus betulinus
Ganoderma adspersum
Grifola frondosa
Coriolus versicolor
Polyporus brumalis

GASTEROMYCETES

Phallus impudicus
Scleroderma citrinum
Lycoperdon pyriforme
L.perlatum

ASCOMYCETES

Xylaria hypoxylon
X.polymorpha
Daldinia concentrica
Hypoxylon fragiforma
Nectria cinnabarina
Ascocoryne sarcoides
Peziza repanda
Aleuria aurantia

Miss Gulielma Lister, a past president of the Essex Field Club recorded the slime-mould *Comatricha tenerrima* from Lords Bushes in March 1939 (the specimen is in the Passmore Edwards Museum). No other recording of this group has been attempted.

THE FAUNA

Epping Forest has for many years been well served by visiting naturalists. Publications on its fauna though, are often either in the form of records of rare species from widely dispersed localities within the Forest or as papers on discrete sections of the fauna such as those by Buck (Coleoptera) (8), Corke (Mammals) (15), Payne (Diptera: Tipulidae), Pinniger (Odonata) and Womersly (Collembola & Thysanura) (52) which cover the whole Forest. There have been few attempts to deal in depth with the fauna of any one area within the Forest. Because of this and also our poor knowledge of a number of groups, (particularly some invertebrate orders) I have produced accounts of the following for Lords Bushes.

Slugs and Snails

Lords Bushes has a very poor mollusc fauna and recording in the woodland is very unrewarding probably due to its acid nature. The list of species is undoubtedly enhanced by recording carried out along the Northern Edge of Lords Bushes; here gardens back on to the woodland along its entire length and from these people dispose of rubble (including old mortar) rubbish (such as rusty old metal and wooden planks) and bundles of garden cuttings and these provide the moist shaded conditions favoured by molluscs. The old mortar no doubt provides the necessary lime, absent in the woodland itself, for the 'growth' of snail shells. In addition to this, a number of species are synanthropic including *Milax budapestensis* and *Helix aspersa*.

All recording was carried out in 1979 from three recording units to Lords Bushes (1) The Undisturbed Woodland, (3) The Burnt and Disturbed Area and (5) The Northern Edge (the numbers in the text below refer to these recording units). Nomenclature follows Kerney, M. P. & Cameron, R. A. D. (36).

Aegopinella nitidula Apparently uncommon, only two empty shells found along (5).
Arion ater Infrequent, found in (1), (3) and (5).
Arion hortensis Probably the commonest of the slugs in Lords Bushes found in (1), (3) and (5).
Capaea nemoralis Found only in (5) and even there is only to be found in the vicinity of the Kings Place entrance to Lords Bushes, though it is to be found in some numbers here (*Cepaea hortensis*, a closely related species, may also be present here but in considerably smaller numbers).
Cochlicopa lubrica Known from only a single specimen from a pile of rubble in (5).
Deroceras reticulatum Frequent, though found only in (5).
Discus rotundatus Infrequent, specimens noted from (1) and (5).
Euconulus alderi Uncommon, only two specimens found beneath fallen beech bark in (1).
Helix aspersa Infrequent, noted only in (5).
Limax maximus Frequent, noted from (1) and (3).
Milax budapestensis Frequent, though noted only from (5).
Oxychilus alliarius Infrequent, noted mainly from (5) I have found three specimens in (1).
Oxychilus cellarius More frequent than O.alliarius and is particularly common along (5) though I have found 3 specimens under a log in (3).
Trichia hispida Frequent, though noted only from (5) under rubble and garden refuse.

Centipedes and Millipedes

The following were recorded from Lords Bushes in 1979. Nomenclature follows Blower, J. G. (5) & Eason, E. H. (20).

Lithobius forficatus I noted very commonly under logs, bark and rubble in most areas of Lords Bushes and along with *Polydesmus angustatus* was common on the burnt and disturbed area. *Blaniulus guttulatus* and *Haplophilus subterraneus* I recorded only from the North Edge of Lords Bushes. I found a number of specimens of *Lithobius melanops* in a well decayed grey squirrel drey built in an oak tree.

CHILOPODA

Necrophloeophagus longicornis
Cryptops hortensis
Haplophilus subterraneus
Lithobius forficatus
melanops

DIPLOPODA

Blaniulus guttulatus
Proteroiulus fuscus
Polydesmus angustatus
Ophyiulus pilosus
Cylindroiulus punctatus

Woodlice

The woodlice (CRUSTACEA: ISOPODA) recorded below are all terrestrial (the common aquatic 'woodlouse' *Asellus aquaticus* occurs in pond P1). The woodlice are probably under-recorded and species such as the minute *Trichoniscus pygmaeus* are almost certainly present. The following species were recorded in 1979. The numbers are the recording areas as described for the flora.

Nomenclature follows Edney, E. B. (21).

Armadillidum vulgare Uncommon, I have found only a single specimen of this species under half a brick on the Forest Edge side of (1).
Oniscus asellus Very common, I find this species over much of Lords Bushes under logs on (3) and noted it to be particularly common in (5) under rubble garden waste and discarded planks of wood.
Philoscia muscorum Common, though I have found only a single specimen on (3) and it is common only in (5) under discarded rubble and garden waste.
Porcellio scaber Common, often noted with *Oniscus asellus*.
Trichoniscus pusillus Probably quite frequent, I found a number of specimens of this small species under a damp well-rotted log in (1).

Butterflies

Lords Bushes, in common with the rest of Epping Forest, is now a poor habitat for butterflies. The total number of species recorded for Epping Forest in recent years is only 25 and of these I have recorded just 16 for Lords Bushes in 1979. There are three further species I would have expected to have recorded: *Lasiommata megera* (Wall Brown), *Cynthia cardui* (Painted Lady), and *Gonepteryx rhamni* (Brimstone), all these species I have noted in the vicinity of Lords Bushes in recent years. The Painted Lady is a migrant species and had 1979 been a 'good' year for this species I would undoubtedly have been able to record it. The Brimstone I have not seen for some years now, the butterflies may have been casual wanderers into this area as the larval foodplant, buckthorn, (*Rhamnus catharticus*) does not grow (as far as I am aware) in the vicinity of Lords Bushes. The Wall Brown's absence may be due to under-recording though I have only seen one or two specimens locally recently.

The following species were recorded for Lords Bushes in 1979. Nomenclature follows Kloet, G. S. & Hincks, W. D. (38).

PIERIDAE
Pieris brassicae (Large White) Noted frequently about Lords Bushes.
Pieris rapae (Small White) Noted frequently about Lords Bushes.
Pieris napi (Green Veined White) Noted frequently about Lords Bushes.
Anthocharis cardamines (Orange tip) A single male noted on (3) on 16th June.

SATYRIDAE
Maniola jurtina (Meadow Brown) Infrequent, including a maximum of 4 or 5 specimens seen at any one time on the burnt and disturbed area.
Coenonympha pamphilus (Small Heath) Infrequent, a few specimens noted on the old grasslands of Lords Bushes.

NYMPHALIDAE
Vanessa atalanta (Red Admiral) A single specimen noted on 5th August on (3).
Inachis io (Peacock) Noted mainly in spring and summer 3 or 4 specimens noted feeding on Buddleia on (3) on 11th August.
Aglais urticae (Small Tortoiseshell) Noted frequently about the burnt and disturbed area.
Polygonia c-album (Comma) Infrequent, noted in spring and late summer including two specimens noted on (3) on 11th September feeding from Michaelmas daisy.

Comma butterfly

LYCAENIDAE
Quercusia quercus (Purple Hairstreak) A single specimen noted flying about some low oak trees in the N.E. corner of Lords Bushes.
Lycaena phlaeas (Small Copper) A single specimen noted on Michaelmas daisy on 24th September on (3).
Celastrina argiolus (Holly Blue) A single specimen noted about holly on the North Edge of the undisturbed woodland on 19th May.
Polyommatus icarus (Common Blue) A single female noted from (3) on 26th August.

HESPERIIDAE
Thymelicus sylvestris (Small Skipper) Infrequent, a maximum of two or three specimens noted on (3) at any one time.
Ochlodes venata (Large Skipper) Infrequent, a single specimen noted on (3).

Two-winged flies

The Diptera, of all the major insect orders, are probably the most neglected in Epping Forest. The only comprehensive work on this group is that of Payne on the Crane-flies (Tipulidae) (52), though there are a number of isolated published records, usually of rare species, taken in the Forest.

In recording this group I have largely confined my collecting to the burnt and disturbed area and to the year 1979. The majority of the species are common and widespread in Britain though there are a number of rare or uncommon species: *Lonchoptera tristis*, *Macrocera maculata*, *Rhamphomyia gibba*, *Systenus pallipes*, *Eustalomyia hilaris*, *Phaonia trigonalis*, *Steganina coleoptrata* and *Leia piffardi*.

Most species recorded are typical of woodland, among them many of the Tipulidae and Mycetophilidae and the robber-flies *Neoitamus cyanurus*, *Dioctria baumhaueri* and *Dioctria linearis*. Some of the hover-flies recorded are also typically woodland species, among them *Ferdinandea cuprea*, a species whose larvae feed in sap-runs, while two species which have been reared from decaying beech wood are *Xylota sylvarum* and *Xylotamima nemorum*. The rare *Systenus pallipes* has been taken on very few occasions in Britain and in common with other members of the genus is more usually reared from rot-holes or sap-runs (elm and beech are recorded) and many of the records of this genus come from long established woodlands.

The burnt and disturbed area with its abundance of grasses, rushes and sedges has provided habitat for a number of grassland species among them members of the Chloropidae. The abundant soft rush (*Juncus effusus*) growing on this area no doubt accounts for the presence in some numbers of *Loxocera albiseta* and *Cerodontha luctuosa*, two species as larvae recorded mining the stems of this plant. Likewise, one or other of the sedges (*Carex* spp.) probably 'hosts' the larvae of *Cordilura albipes*. The fly *Lauxania cylindricornis* is a species typically found in acid grassland (64).

A number of the Diptera recorded are typically found in wet or marshy areas. These include *Anasimyia lineata*, *Epicypta limnophila*, *Platycheirus fulviventris* and *Hydromya dorsalis*, the latter is also of interest since it is the sole representative of the Sciomyzidae taken on the burnt and disturbed area, this no doubt reflects the poor mollusc fauna of this area (Sciomyzid larvae predate molluscs) (64).

It takes many years of recording to even approach finality in the list of Diptera of a given area and finality in such a list is impossible to achieve. Parmenter (48–50) after 25 years of recording on the 380-acre Bookham Common in Surrey listed 1,175 species of Diptera but over 700 of these records were made in the first 3½ years of the survey. It is difficult to be sure of the absence of an insect from an area particularly as I have recorded largely over just a single year (1979) but a number of Diptera have been conspicuous by their absence. The lack of aquatic habitats no doubt accounts for the absence of families such as the Dixidae and Chaoboridae (though the latter at least could occur in the vicinity of the adjacent Knighton Pond), similarly the lack of any large mammals would account for the absence of families such as the Tabanidae. I neither captured nor saw any of the common 'cleg' (*Haematopota pluvialis*) though I took this species elsewhere in Epping Forest in 1979. Other families, the representatives of which are rarely met with, such as the Acroceridae, the Xylomyiidae and some of the uncommon acalypterate families are expected absences. I was, however, surprised not to record any Bombyliidae; I did much work in the early spring in Lords Bushes in 1979 and though noting the bee-fly *Bombylius major* not uncommonly about many grassy rides in Epping Forest I saw no specimens in Lords Bushes.

The majority of Diptera were collected, unless otherwise stated, by using a net 'sweeping' the burnt and disturbed area. The larger insects were tubed individually and the smaller insects collected from the net using a pooter.

Nomenclature follows Kloet, G. S. & Hincks, W. D. (41) with the exception of *Epicypta limnophila* nom.nov.

DIPTERA

NEMATOCERA

TRICHOCERIDAE

Trichocera annulata
hiemalis
saltator

TIPULIDAE

Dictenidia bimaculata
Nephrotoma flavipalpis
quadrifaria
Tipula lunata
marmorata
maxima
obsoleta
oleracea
paludosa

rufina
scripta
Limonia chorea
dumetorum
duplicata
nubeculosa
Pedicia immaculata
Ula sylvatica
Austrolimnophila ochracea
Erioptera hybrida

PSYCHODIDAE

Pericoma sp.

PTYCHOPTERIDAE

Ptychoptera minuta

CULICIDAE

Aedes annulipes
cantans
punctor
Culiseta annulata cantans
Culex pipiens

CERATOPOGONIDAE

Forcipomyia brevipennis
Dasyhelea dufouri

CHIRONOMIDAE

Metriocnemus sp.
Chironomus dorsalis
Psectrotanypus varius

SIMULIDAE

Simulium erythrocephalum

ANISOPODIDAE

Sylvicola cinctus

BIBIONIDAE

Bibio lanigerus
johannis
marci
Dilophus febrilis
femoratus

MYCETOPHILIDAE

Bolitophila tenella
basicornis
Macrocera fasciata
maculata
phalerata
vittata
Orfelia fasciata
Mycomya marginata
Monoclona rufilatera
Apolepthisa subincana
Leia piffardi
Tetragoneura sylvatica
Exechia spinuligera
Phronia johannae
basalis
humeralis
strenua
Dynatosoma fuscicorne
Mycetophila ocellus
trinotata
Sceptonia fumipes
Epicypta limnophila
Platurocypta punctum
testata

SCIARIDAE

Phytosciara flavipes

SCATOPSIDAE

Rhegmoclema sp.

CECIDOMYIIDAE

A few specimens taken but it has not been possible to determine the species.

BRACHYCERA

STRATIOMYIDAE

Beris chalybeata
vallata
Chloromyia formosa
Microchrysa polita
Sargus bipunctatus

RHAGIONIDAE

Rhagio lineola

ASILIDAE

Neoitamus cyanurus
Dioctria baumhaueri
linearis

EMPIDIDAE

Drapetis pusilla
Tachypeza nubila
Platypalpus agilis
clarandra
longicornis
nigritarsis
notatus
pectoralis
Hybos culiciformis
Bicellaria vana
Ocydromia glabricula
Trichonomyia flavipes
Rhamphomyia erythropthalma
gibba
tarsata
Empis aestiva
livida
nigripes
nuntia
praevia
Hilara interstincta
litorea
manicata
Heleodromia immaculata
Dolichocephala guttata
irrorata
Trichopeza longicornis

DOLICHOPODIDAE

Sciapus platypterus
Dolichopus discifer
Hercostomus cupreus
aerosus
Scellus notatus
Medetera sp.
Syntormon denticulatus
Systenus pallipes
Neurigona quadrifasciata
Argyra argyria
Campsicnemus curvipes
scambus

CYCLORRHAPA

ASCHIZA

LONCHOPTERIDAE

Lonchoptera furcata
lutea
tristis

PHORIDAE

A number of specimens taken and awaiting determination.

PLATYPEZIDAE

Opetia nigra
Callomyia amoena
Paraplatypeza atra
Polyporivora picta

PIPUNCULIDAE

Chalarus spurius
Verrallia aucta
Tomosvaryella sylvatica
Pipunculus campestris
Cephalops semifumosus

SYRPHIDAE

Syrphus ribesii
torvus
vitripennis

Metasyrphus corollae
latifasciatus
luniger
Scaeva pyrastri
Dasysyrphus albostriatus
venustus
tricinctus
Melangyna lasiopthalma
cincta
Parasyrphus punctulatus
Xanthogramma pedissequum
Meliscaeva auricollis
cinctella
Episyrphus balteatus
Sphaerophoria scripta
Chrysotoxum cautum
festivum
Baccha elongata
Melanostoma mellinum
scalare
Platycheirus albimanus
angustatus
clypeatus
fulviventris
manicatus
pelatus
scutatus
Pipiza noctiluca
luteitarsis
Pipizella varipes
Neocnemedon vitripennis
Cheilosia grossa
pagana
vernalis
Ferdinandea cuprea
Sphegina kimakowiczi
Neoascia podagrica
Eumerus tuberculatus
Volucella bombylans
pellucens
Xylota segnis
sylvarum
Xylotamima nemorum
Syritta pipiens
Merodon equestris
Helophilus hybridus
parallelus
pendulus
Anasimyia lineata
Parhelophilus versicolor
Eristalis arbustorum
intricarius
nemorum
pertinax
tenax
Eristalinus sepulchralis
Myathropa florea

CONOPIDAE

Conops quadrifasciatus

SCHIZOPHORA

ACALYPTERATAE

TEPHRITIDAE

Phagocarpus purmundus
Orellia ruficauda
Xyphosia miliaria
Sphenella marginata
Paroxyna misella
Tephritis cometa
conjuncta
vespertina

MICROPEZIDAE

Calobata cibaria

PSILIDAE

Loxocera albiseta
Psila rosae

DRYOMYZIDAE

Dryomyza analis
flaveola

LAUXANIIDAE

Minettia inusta
lupulina
longipennis
Sapromyza sordida
Peplomyza litura
Lyciella decipiens
platycephala
rorida
subfasciata
Calliopum aeneum
Lauxania cylindricornis
Tricholauxania praeusta
Homoneura consobrina

HELEOMYZIDAE

Suillia bicolor
affinis
variegata
Heteromyza rotundicornis
Tephroclamys tarsalis

CHYROMYIDAE

Gymnochiromyia inermis

SEPSIDAE

Nemopoda nitidula
Sepsis cynipsea
fulgens
punctum
violacea

SCIOMYZIDAE

Elgiva cucularia
Hydromya dorsalis

SPHAEROCERIDAE

Sphaerocera curvipes
Copromyza atra
fimetaria
stercoraria
uncinata
Leptocera appendiculata
clunipes
claviventris
coxata
curvinervis
humida
ochripes
rufilabris

PALLOPTERIDAE

Palloptera umbellatarum

LONCHAEIDAE

Lonchaea contigua
laticornis
scutellaris
sylvatica

PIOPHILIDAE

Piophila vulgaris

OPOMYZIDAE

Geomyza balachowskyi
tripunctata
Opomyza florum
germinationis

CLUSIIDAE

Clusia flava

ANTHOMYZIDAE

Anthomyza sordidella

ASTEIIDAE

Asteia amoena
Leiomyza scatophagina

CAMILLIDAE

Camilla sp. (nigrifrons?)

EPHYDRIDAE

Dichaeta caudata
Ilythea spilota
Philygria stictica

Hyadina guttata
Limnellia quadrata
surturi
Coenia curvicauda
pulustris
Paracoenia fumosa

DIASTATIDAE

Diastata fuscula

DROSOPHILIDAE

Steganina coleoptrata
Leucophenga maculata
Scaptomyza flava
graminum
pallida
Drosophila andalusiaca
cameraria
hydei
immigrans
melanogaster
obscura
subobscura

AGROMYZIDAE

Agromyza mobilis
rondensis
Liriomyza flaveola
Phytomyza ilicis
nigra
ranunculi
Cerodontha denticornis
luctuosa

CHLOROPIDAE

Oscinella frit
maura
nitidissima
Elachiptera cornuta
diastema
tuberculifera

SCHIZOPHORA

CALYPTRATAE

TACHINIDAE

Eriothrix rufomaculata
Tachina fera
Carcelia lucorum
Triarthria spinnipennis
Pales pavida
Phryxe nemea

SARCOPHAGIDAE

Metopia argyrocephala
Brachicoma devia
Sarcophaga carnaria
crassimargo
incisolobata
melanura
subvicina

CALLIPHORIDAE

Calliphora vicina
vomitoria
Cynomya mortuorum
Lucilia caesar
Pollenia rudis

SCATHOPHAGIDAE

Norellisoma spinimanum
Cordilura albipes
Nanna fasciata
inermis
Cleigastra apicalis
Scathophaga furcata
lutaria
stercoraria

ANTHOMYIIDAE

Pegohylemyia fugax
Lasiomma meadi
Hydrophoria linogrisea
Chelisia monilis
Eustalomyia hilaris
Delia platura
Paregle radicum
Nupedia aestiva

FANNIIDAE

Fannia armata
coracina
fuscula
genualis
hamata
serna

MUSCIDAE

Mesembrina meridiana
Dasyphora cyanella
Orthellia viridis
Morellia simplex
Musca autumnalis
Azelia macquarti
Hydrotaea dentipes
irritans
Muscina pabulorum
Phaonia basalis
errans
fuscata
halterata
pallida
palpata
populi
signata
trigonalis
variegata
Helina depuncta
impuncta
laetifica
Gymnodia humilis
Hebecnema affinis
vespertina
Mydaea ancilla
Spilogona vana
Lispocephala alma
erythrocera
Schoenomyza littorella
Allognota agromyzina

Bees, Wasps and Ants

The Hymenoptera are another group very much under-recorded not only for Epping Forest but in the county of Essex as a whole. They are the largest order of British insects, though the majority of species are found in just one group, the Parasitica, which are mostly small to minute insects such as those found in the super-family Chalcidoidea (with over 1,500 species in Britain alone) whose taxonomy is therefore very difficult. I have restricted my recording to the Aculeata which includes the familiar bees and wasps. The only historical record I can find is of the hornet (*Vespa crabro*) which was noted nesting in Lords Bushes in the 1940s (69).

The burnt and disturbed area has been readily colonised by many species of bee and wasp. I have found the nest of the common wasp (*Vespula vulgaris*) in the decaying interior of the stump of a felled oak pollard and in September 1977 I noted a number of specimens of the four-spotted digger (*Crossocerus quadrimaculatus*) excavating burrows in the sands and gravels caught up in the root system of an upturned birch stump. Other species probably making use of the tree stumps and decaying wood include the mining bee (*Anthophora furcata*) and the leaf cutter bee (*Megachile willughbiella*). A number of species are known to be ground nesters, among them the common tawny mining bee (*Andrena fulva*). I noted only a solitary female of this species in Lords Bushes though often they occur in discrete colonies (there is a very strong colony of this species on grassland opposite the Roebuck Inn). Another common mining bee (*Osmia rufa*) I noted collecting particles of damp clay from the side of a ditch, presumably to make nest cells with. The familiar bumble-bees (*Bombus* spp.) were well represented including the apparently uncommon (in Essex) red-shanked carder bee (*Bombus ruderarius*) though this is almost certainly due to under-recording. I was surprised to record 5 of the 6 British species of cuckoo bee (*Psithyrus* spp.) on the burnt and disturbed area, these are brood parasites of bumble-bees; all the required host species were recorded but it is not always the case that the parasite is recorded as well. The homeless bees (in Lords Bushes I have recorded only the common *Nomada goodeniana*) are also brood parasites, though of genera such as *Andrena* and *Halictus*. Also on the topic of parasites, I was surprised, given in particular the number of *Bombus* spp. present to record just a solitary specimen of thick-headed fly *Conops quadrifasciatus* from the burnt and disturbed area. Two uncommon bees noted were *Lasioglossum pauxillum* and *Sphecodes puncticeps*.

The following species were recorded mainly in 1979. Nomenclature follows Kloet, G. S. & Hincks, W. D. (40).

FORMICOIDEA

Myrmica rubra
Lasius niger

VESPOIDEA

Dolichovespula sylvestris
Vespula germanica
rufa
vulgaris

POMPILOIDEA

Dipogon variegatus

SPHECOIDEA

Argogorytes mystaceus
Crossocerus annulipes
varus
quadrimaculatus
Ectemnius cavifrons
Psen dahlbomi

APOIDEA

Andrena bicolor
fulva
Anthophora furcata
Apis mellifera
Bombus hortorum
lapidarius
lucorum
pascuorum
pratorum
ruderarius
terrestris
Halictus tumulorum
Hylaeus communis
Lasioglossum pauxillum
villosulum
Megachile circumcincta
willughbiella
Nomada goodeniana
Osmia rufa
Psithyrus barbutellus
campestris
rupestris
sylvestris
vestalis
Sphecodes puncticeps

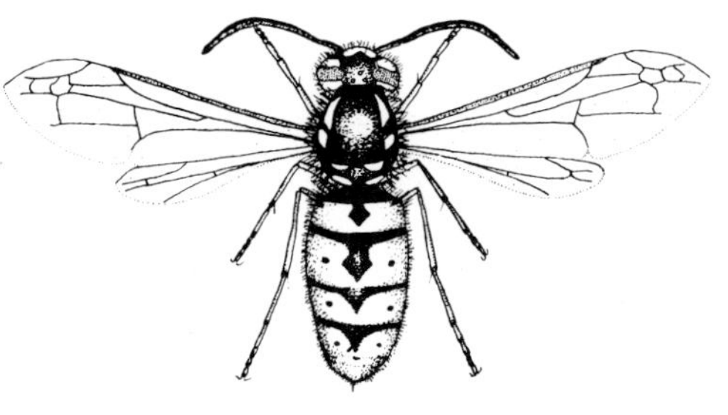

German Wasp (*Vespula germanica*)

Miscellaneous insect records

The insects listed below are mainly the result of casual collecting on the burnt and disturbed area of Lords Bushes in 1979–1982. Only the Odonata and Orthoptera are anywhere near complete (i.e. the theoretical maximum number of species the habitat could support). The ponds available in Lords Bushes would appear to be unsuitable for dragonfly nymphs and all the species recorded are likely to have wandered from the nearby Knighton Pond. With regard to the number of species of Orthoptera recorded (just two species despite exhaustive and concentrated search), I think it likely that insufficient time has elapsed since the burnt and disturbed area was created, after the events of 1976, for other species of Orthoptera to colonise. Prior to 1976 Lords Bushes had 97% tree cover and was thus an unsuitable habitat for Orthoptera.

Nomenclature for these groups follows Kloet, G. S. & Hincks, W. D. (37–40).

ODONATA

Aeshna cyanea
grandis
Enallagma cyathigerum
Ischnura elegans
Lestes sponsa
Sympetrum striolatrum

ORTHOPTERA

Chorthippus brunneus
Meconema thalassinum

DERMAPTERA

Forficula auricularia

HEMIPTERA

Tuberolachnus salignus

NEUROPTERA

Chrysopa carnea
perla
Hemerobius humulinus
micans
Raphidia xanthostigma

LEPIDOPTERA

Agriphila tristella
Anthophila fabriciana
Blastobasis lignea
Coleophora alticolella
Crambus pascuellus
Cydia splendana
Endotricha flammealis
Epagoge grotiana
Gypsonoma dealbana
Mompha propinquella
raschkiella
subbistrigella
Myelois cribrella
Olethreutes lacunana
Plutella xylostella

HYMENOPTERA

Andricus fecundator
kollari
quercuscalicis
Biorrhiza pallida
Neuroterus numismalis
quercusbaccarum

COLEOPTERA

CARABIDAE

Abax parallelopipedus
Agonum albipes
Amara anthobia
plebeja
Bembidion harpaloides
lampros
lunulatum
quadrimaculatum
tetracolum
Loricera pilicornis
Nebria brevicollis
Notiphilus biguttatus
Pterostichus madidus
nigrita
strenuus

Others

Clytus arietis
Cylindronotus laevioctostriatus
Glischrochilus hortensis
Platypus cylindrus
Pogonocherus hispidus
Strangalia maculata
Xyloterus domesticum

William Cole, founder of the Essex Field Club (who lived at Buckhurst Hill) provided the first beetle record for Lords Bushes: the rare chafer *Gnorimus nobilis* taken in July 1881 (67). He also recorded the huge 'longhorn' beetle *Prionus coriarius* from the Buckhurst Hill area (9). The species listed above were all taken on the burnt and disturbed area of Lords Bushes in 1979 and 1980 with the exception of *Xyloterus domesticus* (which I reared in some numbers from Beech logs). Whilst collecting I concentrated on ground beetles (Carabidae) and these are listed separately.

Amphibians and Reptiles

Fortunately, due to the work of David Scott (pers. comm.), mainly in the 1950s, the account of the amphibian and reptile fauna of Lords Bushes is probably complete. During my recording (mainly in 1979) it has become obvious that there has been some decline in the species present and it is likely that two factors are involved: (a) the ponds P1 and P2 have become infilled with beech leaves and also the dam was broken for the pond P2 and hence they dry out in summer making them unsuitable habitats for amphibian young and (b) Lords Bushes had become very overgrown with excessive overshading by the beech trees and consequently unsuitable habitat for the sun loving reptiles. No reptiles and only two amphibians were recorded in the 1979 survey.

Crested Newt *Triturus cristatus*
This, the largest of the three British species of newt, was recorded by Scott from ponds P1 and P2 in the years 1953–59 in 'fair numbers' though with a preference for P1, which was deeper and more shaded than P2. This includes records in 1953 of this species breeding in car tyres discarded in P1. Not recorded on the 1979 survey.

Smooth Newt *Triturus vulgaris*
This species was recorded as being common in the ponds in Lords Bushes in the years 1953–59. I have not recorded this species in Lords Bushes in 1979 but it is abundant in my garden pond (adjacent to Lords Bushes) and, including lesser numbers of the palmate newt, I have counted in one night in excess of 60 newts in the pond which is about 3.5m across.

Palmate Newt *Triturus helveticus*
This was the only species of newt I recorded in Lords Bushes in 1979, noted under logs and on a number of occasions under rubble in (5) the North Edge of Lords Bushes. Scott noted them to be common from 1953–59 in the ponds and noted they were more frequently encountered than the smooth newt.

Common Toad *Bufo bufo*
Described as the most numerous amphibian in Epping Forest by Malenoir (43) most records of this species in Lords Bushes are of animals migrating from their winter hibernating quarters (probably local gardens) to their breeding grounds, Knighton Pond. Toads were noted in some numbers in 1953 and on 26 March 1956 approximately 70 were noted moving towards Knighton Pond. I have no record of this species for 1979 in Lords Bushes, though in all probability it still breeds in Knighton Pond.

Common Frog *Rana temporaria*
Scott noted this species to be fairly common in the Lords Bushes ponds in 1953–59 and previously in the late 1940s he had recorded as many as 80 pairs of this species in the pond P2 and in the 1960s noting a decline which is still evident. I recorded spawn from only 3 or 4 females in 1979 deposited in one shallow corner of the pond. Young frogs and the occasional adult I noted on the burnt and disturbed area.

Slow Worm *Anguis fragilis*
This secretive lizard was recorded by Scott between 1950–55 on waste land bordering Lords Bushes. It is often found in open habitats, where it is able to bask, and this would seem to preclude its occurrence in the wooded area of Lords Bushes. This species still occurs in gardens which back on to the Buckhurst Hill/Woodford branch of the Central Line where it is (along with the common lizard) apparently to be found quite commonly, spending the winter months hibernating in compost heaps.

Grass Snake *Natrix natrix*
Scott recorded 2 or 3 grass snakes 'near Forest Edge' about 1950–54. I have not recorded this species on the 1979 survey.

Adder *Vipera berus*
Adders were certainly present in Lords Bushes up to 1957 and probably until the mid 1960s, when I recall finding a young one (which I reluctantly had to destroy) in a neighbour's garden adjacent to Lords Bushes.

Birds

The history of birds in Lords Bushes is not well documented. Buxton (9) mentions a number of species that occurred on and about his estate (Knighton, adjacent to Lords Bushes) and other species then common in Epping Forest, which are likely to have occurred in Lords Bushes. Among the records of particular interest are sparrowhawk, which bred at Knighton, and the barn owl which Buxton noted in Lords Bushes. Other records include the hawfinch which was recorded in Lords Bushes in 1933 and the rather startling record of redshank, a pair of which nested in a field adjacent to Lords Bushes in 1932 (4).

The present recording period covers the years 1978–1980 with most records from 1979–1980. The majority of the species listed below are typical of lowland mixed deciduous woodland. The burnt and disturbed area has obviously attracted a number of species, among them seed-eaters such as the redpoll, goldfinch, and reed bunting in the winter months. In summer, the dense shrubby growth and open areas adjacent to woodland, make it a good habitat for warblers. It is interesting to speculate that recently felled woodland would be a suitable habitat for the nightjar, though for a variety of reasons, among them a drastically declining national population and, locally, probably what would be excessive human disturbance, I have not been able to record this species for Lords Bushes.

Mallard *Anas platyrhynchos*
Can be seen in small numbers at any time of the year including a maximum of 21 (from the adjacent Knighton Pond) feeding on the burnt and disturbed area. On the 27th April 1979 a duck with 2 ducklings was noted on the bomb crater C1.

Kestrel *Falco tinnuculus*
I noted a male bird frequently in the winter months about the burnt and disturbed area in search of prey. This species was recorded in 1956 (Scott pers. comm.) nesting in a beech tree in Lords Bushes.

Moorhen *Gallinula chloropus*
Seen only rarely on the two ponds P1 and P2, though it breeds on nearby Knighton Pond.

Woodcock *Scolopax rusticola*
A single bird seen on the burnt and disturbed area on 8th Nov. 1980.

Stock Dove *Columba oenas*
As many as 3 pairs of this species may have bred in Lords Bushes in 1979, a pair were noted inspecting a nest hole in a beech tree as early as the 8th March.

Wood Pigeon *Columba palumbus*
Present as a breeding bird and was noted often in some numbers (max. 180 birds) in the winter of 1978–79.

Feral Pigeon *Columba livia*
Small flocks of this species can be seen at most times of the year, usually about the margin of Lords Bushes.

Collared Dove *Streptopelia decaocto*
Can be seen, though infrequently, at any time of the year usually in small numbers (one or two birds).

Cuckoo *Cuculus canorus*
More often heard than seen, my first record for Lords Bushes in 1979 was on the 27th May, probably a single bird spent the summer of 1980 in Lords Bushes.

Tawny Owl *Strix aluco*
Present all year round and a pair is probably breeding, occasionally seen out in daylight mobbed by jays, magpies, and small birds.

Green Woodpecker *Picus viridis*
Noted usually as a single bird and perhaps more frequently in autumn and winter, though I have records for this species throughout the year.

Great Spotted Woodpecker *Dendrocopus major*
Noted mainly in the winter months with a maximum of two birds seen, though I have also heard and watched this species drumming and it probably breeds in Lords Bushes.

Great Spotted Woodpecker

Lesser Spotted Woodpecker *Dendrocopus minor*
I first noted this inconspicuous little bird on the 1st April 1979 in Lords Bushes and one and possibly two birds were present at least until the 14th May.

Carrion Crow *Corvus corone*
Maximum of usually four or five birds present in the winter months and one pair has certainly bred, nesting in a tall beech tree.

Jackdaw *Corvus monedula*
Noted very infrequently over and about Lords Bushes, mainly in the winter months.

Jay *Garrulus glandarius*
Present all year round and probably breeds in Lords Bushes, maximum of eight birds noted in December 1978.

Magpie *Pica pica*
Present all year round and a pair has bred in Lords Bushes, four birds were noted in the winter of 1978–79.

Great Tit *Parus major*
Common, present all year round.

Blue Tit *Parus caeruleus*
Common, present all year round.

Coal Tit *Parus ater*
Infrequent, though can be seen in small numbers at any time of the year.

Marsh Tit *Parus palustris*
Infrequent, perhaps more often noted in the winter months. I have a single possible record of the Willow Tit (*P. montanus*) but the presence of this species requires confirmation.

Long Tailed Tit *Aegithalos caudatus*
Noted mainly as family parties in the winter months; a pair nested in 1980 in the regrowth of a felled holly tree on the burnt and disturbed area, though the nest was subsequently destroyed.

Nuthatch *Sitta europaea*
This loquacious little bird is present all year round in Lords Bushes, the maximum number I have counted is five and it is very likely that this species is breeding.

Tree Creeper *Certhia familiaris*
Noted infrequently, usually as single birds, though on the 9th April 1979 I watched two birds chasing one another around the branches of an oak pollard.

Wren *Trogolodytes trogolodytes*
This species is not uncommon and is to be found all year round. I have found two nests of this species neatly constructed in the trunk folds of hornbeam pollards and noted another being built in the underside of the root system of a fallen oak tree.

Mistle Thrush *Turdus viscivorus*
Frequently noted in ones and twos, occasionally more, about Lords Bushes. A pair nested in 1980 on the bough of a beech tree.

Fieldfare *Turdus pilaris*
I have noted this winter visitor only infrequently and in small numbers, perhaps due to the lack of berries on bushes such as holly and hawthorn, though I have seen flocks of up to 30 birds fly over.

Song Thrush *Turdus philomelos*
Not common, but present all year round and certainly breeds in small numbers.

Redwing *Turdus iliacus*
Marginally more frequent than the fieldfare in the winter of 1978–79, with a maximum of four seen on 29th January. Larger flocks fly over (including I believe at night).

Blackbird *Turdus merula*
By far the most frequent of the thrushes present all year round, certainly breeds and I have noted loose flocks of in excess of 20 birds feeding on the burnt and disturbed area.

Robin *Erithacus rubecula*
Frequent, present all year round.

Blackcap *Sylvia atricapilla*
At least two pairs held territories on the edge of the burnt and disturbed area in 1980.

Garden Warbler *Sylvia borin*
Two birds, including a singing male, were noted going in and out of a dense patch of bramble and sallow saplings on the burnt and disturbed area on the 24th May 1980, though subsequently nothing more was seen or heard of the birds.

Chiffchaff *Phylloscopus collybita*
A maximum of six males heard singing about the burnt and disturbed area in 1980.

Willow Warbler *Phylloscopus trochilus*
A maximum of five males heard singing about the burnt and disturbed area in 1980.

Goldcrest *Regulus regulus*
Infrequent, mainly a winter visitor though I have records of this little warbler as late as 9th April.

Spotted Flycatcher *Muscicapa striata*
A few birds seen about the burnt and disturbed area in the spring and autumn; possibly one or two pairs breed.

Dunnock *Prunella modularis*
Frequent, present all year round.

Starling *Sturnus vulgaris*
Present all year round and a very common breeding species.

Tree Pipit *Anthus trivialis*
A single bird noted on the burnt and disturbed area on the 8th June 1980. Possibly a late passage bird, attracted by the cleared woodland, this species is more usually to be seen in the Long Running and Sunshine Plain areas of Epping Forest.

Pied Wagtail *Motacilla alba*
Noted on a few occasions in and about Lords Bushes mainly in the winter months.

Greenfinch *Carduelis chloris*
Again perhaps more frequently noted in the winter months in small flocks.

Redpoll *Acanthis flammea*
Present in the winter months though surprisingly I have records as late as the 3rd May. A flock of 8 or 9 noted feeding on the burnt and disturbed area on the 8th November. I have noted this species feeding on birch seed.

Bullfinch *Pyrrhula pyrrhula*
Infrequent but can be seen in ones and twos (often pairs), occasionally more, in the winter months. A pair nested on the burnt and disturbed area in 1980.

Chaffinch *Fringilla coelobs*
Perhaps the commonest of the finches in 1979, a flock of 40–50 were noted on the 29th November.

Brambling *Fringilla montifringilla*
A number of birds perhaps 8 or 9 were noted feeding on beech mast amongst a flock of chaffinches in December 1979.

Reed Bunting *Emberiza schoeniculus*
A small party of four birds were noted throughout February and March 1979 feeding about the burnt and disturbed area and again two birds were noted about this area in December 1980.

House Sparrow *Passer domesticus*
Frequently noted in and about Lords Bushes.

In addition to the above species, the following birds were noted flying over or near Lords Bushes in 1979 and 1980:
Ardea cinerea (Grey Heron), *Vanellus vanellus* (Lapwing), *Larus argentatus* (Herring Gull), *L. canus* (Common Gull), *L.fuscus* (Lesser Black Back), *L. Marinus* (Greater Black Back), *L.ridibundus* (Black Headed Gull). Gulls have a regular diurnal pattern of movement over Lords Bushes in the autumn and winter months; as daylight fades flocks totalling up to several hundred birds come in, I presume, from feeding in ploughed fields, rubbish tips and sewage works in Essex, probably making for the Lea Valley resevoirs to roost at night. *Alauda arvensis* (Skylark) noted in large flocks over Lords Bushes in January 1979, as many as 40–50 being noted. *Delichon urbica* (House Martin), *Hirundo rustica* (Swallow), *Psittacula krameri* (Ring Necked Parakeet) noted infrequently over Lords Bushes, usually one to three birds.

Mammals

Lords Bushes is not rich in mammals, just five species were recorded on the 1979 survey, additionally there are records of the house mouse (*Mus musculus*) and the brown rat (*Rattus norvegicus*) from the houses and gardens which abut Lords Bushes. A small mammal trapping survey was carried out mainly on the burnt and disturbed area, but surprisingly no records of bank vole (*Clethrionomys glareolus*) were made nor were any shrews taken but I have heard the latter (probably the common shrew (*Sorex araneus*) on one occasion 'squeaking' during the day. There are possibly a number of genuine absences, the soil is probably too shallow and gravelly for the mole (*Talpa europaea*) which can often be seen in small woodlands. The urban area which surrounds Lords

Bushes with its associated human disturbance must preclude deer and hare (*Lepus europaeus*) though a muntjac (*Muntiacus reevesi*) was killed on the High Road at Buckhurst Hill in 1973 (10). Daniel in his 'Rural Sports' published in 1812 records the inn-keeper of the Bald Faced Stag keeping a tame pine marten (*Martes martes*) and letting it have the run of the kitchens there! (17). At that time the pine marten was a native of Epping Forest. As I have already mentioned rabbits (*Oryctolagus cuniculus*) were formerly present in Lords Bushes and the red squirrel (*Sciurus vulgaris*) was last recorded for Lords Bushes in about 1949 by Scott (pers.comm.). H. A. Cole recorded the dormouse (*Muscardinus avellanarius*) from Buckhurst Hill in 1887. At this time this animal was not uncommon in Epping Forest (12).

Hedgehog *Erinaceus europaeus*
Hedgehogs were noted to be common in the Buckhurst Hill area (including Lords Bushes) in the 1950's by David Scott and this appears to be their status today. Hedgehogs are frequently seen and heard in the Knighton Lane area and I also have a record of a dead hedgehog on the burnt and disturbed area on the 5th December 1978. I have also taken the hedgehog flea (*Archaeopsylla erinacei*) locally.

Pipistrelle *Pipistrellus pipistrellus*
This little bat can be seen late on most summer evenings in Lords Bushes hawking around the clearings, rides and over the ponds for insects. A few years ago (I have no evidence for its existence in 1979) there was a fairly large roost of this species in the loft of one of the relatively newly built houses in Wimborne Close, access being gained by a slipped roof tile.

Grey Squirrel *Sciurus carolinensis*
The grey squirrel is very common in Lords Bushes, the maximum number I have counted in the woods in winter on a warm sunny day is 18 on the 10th January 1979, but this is probably an underestimate of the total population. In winter this species seems to gather in small groups, I have recorded as many as 10 together, possibly for mating which takes place in the winter months. I have also taken the squirrel flea (*Orchopeas howardi*) locally.

Wood Mouse *Apodemus sylvaticus*
A small mammal trapping survey of 42 trap-nights produced 19 small mammals, all of this species (none of these examined proved to be the yellow necked mouse (*Apodemus flavicollis*). Trapping was mainly carried out in the burnt and disturbed area (36 of the 42 trap nights) which produced 18 of the 19 captures of wood mice.

Red Fox *Vulpes vulpes*
This species is present though it is difficult to ascertain numbers. An adult and two cubs were noted on the main path on the 30th October 1978. There is a disused earth under a garden shed in a garden abutting Lords Bushes. I have found 'scats' and noted foxes on the burnt and disturbed area on several occasions.

DISCUSSION

I hope I have shown how changes in habitat, due to changes in management practice, can occur over the decades and centuries. In 1976 few people would have realised that Lords Bushes once contained a heathland element but an examination of the flora that has appeared following the clearances would make this appear to be so.

What has happened in Lords Bushes is a microcosm of the events that have happened over much of Epping Forest. When managed as wood pasture the Forest contained areas of pollard trees, discrete grassy plains, areas of heathland, marshes and bogs. The cessation of pollarding and the subsequent removal of many pollards and the encouragement of standard trees has resulted in dense, floristically poor and historically uninteresting, woodland with many of the remaining pollards being overshadowed and killed by the more vigorous standard trees. The decline of grazing has enabled coarse grasses and scrub to invade the grassy plains and birch the bogs and heathy areas. The net result of the cessation of management as wood-pasture is a decline in the diversity of habitat with a concomitant decline in the diversity of the flora and fauna (14).

It must be no coincidence that many of the plants listed below, typical of heathland, heathy commons and acid bogs present in the 19th century in Epping Forest, are now extinct (E) or rare and declining (D) in the Forest (34).

Bog Pimpernel	*Anagallis tenella* (E)
Ribbed Sedge	*Carex binervis* (D)
Straight-beaked Sedge	*demissa* (D)
Flea Sedge	*pulicaris* (E)
Meadow Thistle	*Cirsium dissectum* (E)?
Bell Heather	*Erica cinerea* (D)
Cross-leaved Heath	*tetralix* (D)
Cotton Grass	*Eriophorum angustifolium* (D)
Needle Whin	*Genista anglica* (D)
Pennywort	*Hydrocotyle vulgaris* (D)
Heath Rush	*Juncus squarrosus* (D)
Blunt-flowered Rush	*subnodulosus* (E)
Common Clubmoss	*Lycopodium clavatum* (E)
Marsh Clubmoss	*inundatum* (E)
Lousewort	*Pedicularis sylvatica* (D)
Lesser Butterfly Orchid	*Platanthera bifolia* (E)
Heath Milkwort	*Polygala serpyllifolia* (D)
Trailing Tormentil	*Potentilla anglica* (E)
Shepherd's Cress	*Teesdalia nudicaulis* (E)
Dwarf Furze	*Ulex minor* (D)
Ivy-leaved Bell-flower	*Wahlenbergia hederacea* (E)

A classic example of declining heathland is to be found in Wanstead Flats. As far back as 1199 Wanstead and Leyton Flats are recorded as heath ("bruerio") upon which the monks of Stratford were allowed to send 960 sheep (14) until the 19th century it is clear a rich heathland flora existed here. Joseph Freeman in his "Stratford Flora" of 1862 records plants "to be found within a short walk of Stratford" (28), (an examination of herbarium material in the Passmore Edwards Museum reveals many of these records refer specifically to Wanstead), such as dodder (*Cuscuta epithymum*) on ling (*Calluna vulgaris*), birdsfoot vetch (*Ornithopus perpusillus*), lesser butterfly orchid and dwarf furze. Further examination of herbarium material reveals bell heather (*Erica cinerea*) collected from Wanstead Flats in 1877 by William Cole. Gibson writing in 1862 (29) notes lemon-scented Fern (*Thelypteris limbosperma*) from Wanstead. Today only a fragment of this rich heathland flora remains such species including ling, heath rush and heath grass (*Sieglingia decumbens*). Nearby Leyton Flats however retains much of its heathy character with species such as carnation-grass (*Carex panicea*), oval sedge (*C. ovalis*) heath rush, heath woodrush (*Luzula multiflora*), mat-grass (*Nardus stricta*), creeping willow (*Salix repens*) and heath grass.

Similarly once a particular habitat disappears the associated fauna goes as well, the insects listed below typically found on heaths and heathy commons have in most cases not been recorded in Epping Forest since the 19th or early 20th century (15).

ODONATA
Sympetrum scoticum
DIPTERA
Asilus crabroniformis
COLEOPTERA
Agonum sexpunctatum
Cassida hemisphaerica
Trachys trogolodytes

LEPIDOPTERA
Boloria selene (Small Pearl-bordered Fritillary)
Coleophora potentillae
Diacrisia sannio (Clouded Buff)
Euphydryas aurinia (Marsh Fritillary)
Oncocera palumbella
Plebejus argus (Silver-studded Blue)

Vertebrates too are similarly affected, there are few recent sightings of adders in the Forest (15) possibly due to the decreasing amount of heath and grassy plain available to them. Henry Doubleday, the great Epping naturalist, recorded the stonechat, a classic bird of gorse-clad heaths (30) as being periodically common in Epping Forest (11), today it no longer breeds and is to be seen only as an uncommon winter visitor. The hobby, another bird often associated with heathland is recorded as nesting in the Epping Forest area in 1846-7. Lapwings were recorded nesting on Fairmead Bottom and near the Wake Arms in the 1860's: evidence that these areas were considerably more open and extensive than they are today (15).

The old pollard trees, that other great legacy of the wood-pasture tradition in Epping Forest, are an equally valuable habitat. Although atmospheric pollution has removed many of the characteristic lichens, pollards, are still an important micro-habitat for invertebrates. Numerous species of beetle are known to be associated with the dead and decaying wood (15) and many flies also rely on this micro-habitat (64). Rearing from rot-holes in old beech pollards have provided records of the following rare or uncommon Diptera from Epping Forest (most are associated with ancient woodland) the Dolichopodids *Systenus leucurus* and *S. scholtzi*, the magnificent crane-fly *Ctenophora pectinicornis* and the superb bee-mimicing hover-fly *Pocota personata*; other flies associated with rot-holes and decaying wood and taken in the Forest include *Mallota cimbiciformis, Myolepta luteola, Phaonia trigonalis, Systenus pallipes, Xylomyia maculata, Xylota abiens, Xylotamima lenta and X. nemorum*.

Pollards also provide a favoured nesting site for the redstart which may be a classic bird of working wood-pasture, it favours big mature trees in open areas. Today much of Epping Forest would be unsuitable habitat for this species, dense beech woodland casts too deep a shade and lacking a diverse flora also provides little in the way of insect food.

Marshy areas and bogs have also suffered in Epping Forest. Connaught Waters, was made on the site of a huge marsh shortly after the Epping Forest Act was passed. Sadly, I can find no record of the flora which existed there. Deliberate drainage, invasion by unchecked birch scrub and overshadowing has also contributed to the decline of this valuable habitat in Epping Forest.

Superimposed on the consequences of a disused system of management, and an unsympathetic modern one, are other deleterious factors: increasing pollution (the bryophyte and particularly the lichen flora are greatly impoverished); large numbers of visitors at weekends and other times of the year often to a few over-used areas (High Beach, Wanstead Flats and Connaught Waters for example) can cause problems at these sites. Changes in climate too can be deleterious, it is thought this may be the deciding factor in the dramatically reduced British summering populations of the red-backed shrike and nightjar, both were not uncommon breeding species in Epping Forest at the turn of the century.

A Forest with all its wood-pasture features intact is just as much an historic monument as for example is any ancient building and equally deserving of the highest conservation priority. In Epping Forest we have tangible evidence of a way of life that until the 19th century had probably little altered in over 900 years. Since then, these factors have brought about quite a fundamental change to the ecology of Epping Forest resulting in uninteresting beech woodland with much reduced grassland, heath and bog.

The builders of the A.11 Epping New Road in the 1830's could not have foreseen the effect the invention of motorised transport would have on grazing in the Forest. It is now impossible for Epping Forest to become working wood-pasture once again but I believe it would be greatly beneficial if, in selected areas of manageable size, the old wood-pasture traditions (pollarding and grazing or its modern mechanical equivalent) could be reintroduced. This would certainly benefit the flora and fauna and at the same time be a very valuable educational asset. The events in recent years in Lords Bushes have shown that not only is it feasible but could be carried out with very successful results.

CONCLUSION

Many people would regard the events after 1976 in Lords Bushes as tragic: 'another few hundred of the ever diminishing number of trees lost to the British countryside' but, as I hope I have shown, Lords Bushes and indeed Epping Forest are suffering from too many trees (of the wrong type) with a subsequent loss in the diversity of habitat, a loss of many of the ancient historic features, the plains, heaths and pollards and a decline or outright loss of the associated flora and fauna. Lords Bushes has changed from an overmature floristically poor woodland to one which now maintains a healthy diversity of plant species and a correspondingly richer fauna. Perhaps one can regard this as a restoration of a heathy plain that existed in Lords Bushes many decades ago. In the absence of grazing pressure this plain, like all the others, will disappear; merely a transient phase one of several 'seres' it will go through until in a hundred years or more it is a beech woodland once again. Some attempt will, I hope, be made to arrest this process.

I fervently hope the 'Regents Park Atmosphere' described by the author of the extract reproduced in the editorial of this booklet never materialises!

SUGGESTIONS FOR FURTHER WORK

BIOLOGICAL RECORDING

a. It would be most useful if groups of plants and animals not, or incompletely, worked on the present survey (such as the Coleoptera, Lepidoptera, Araneae or the Fungi) could be recorded.
b. A common bird census should be undertaken on part of the burnt and disturbed area.
c. The changing flora and fauna of the woodland should be monitored.
d. An analysis of the contemporary pollen rain could be undertaken and compared with that recorded after the lime decline in Epping Forest.
e. A study of the annual rings of felled trees should be made.

MANAGEMENT

a. Some parts of the burnt and disturbed area to be kept open to maintain the diversity of habitat within the wood.
b. The pond P2 to be cleared of accumulated debris.
c. The area around the Pulpit Oak to be cleared of encroaching holly and other vegetation.
d. Ditching operations in the wood to cease and if possible ditches to be refilled to prevent excessive drainage.
e. As many as possible of the fallen dead trees be allowed to decay where they lie and not be burnt or removed.
f. Perhaps pollarding could once again be put into operation.

ACKNOWLEDGEMENTS

I would like to thank the Conservators of Epping Forest who granted permission to record, and collect where necesary, the flora and fauna. I would also like to thank Dr Oliver Rackham who provided many valuable comments and criticisms on the original draft of this manuscript; the Essex Field Club who agreed to publish this booklet and also for the loan of Longworth small mammal traps and the following who have been so forthcoming with determinations or confirmation of my determinations:-
Vascular Plants — Dr K. J. Adams
Bryophytes — Dr K. J. Adams
Lichens — Mr J. Skinner (Southend Museum)
Fungi — Mr G. Kibby
Mollusca — Mr F. Naggs (British Museum, Natural History)
Chilopoda & Diplopoda — Mr P. Hillyard (B.M., N.H.)
Isopoda — Ms J. Ellis (B.M., N.H.)
Diptera — Mr P. J. Chandler; Mr J. Dear (B.M., N.H.); Mr A. Irwin (Castle Museum, Norwich); Mr R. Lane (B.M., N.H.); Mr A. Pont (B.M., N.H.); Mr D. A. Smith; Mr A. Stubbs (Nature Conservancy Council); Mr R. Uffen.
Hymenoptera — Mr M. Edwards
Coleoptera — Mr P. M. Hammond
Micro-Lepidoptera — Lt Col. A. M. Emmet

I would particularly like to thank Mr Derek Smith, the Essex Field Club Diptera recorder, for much hard work put into the list of flies; Dr Ken Adams for his numerous identifications and checking of a number of plant species; Mr Geoffrey Kibby who contributed the section on fungi and Mr M. Edwards who determined the smaller Aculeate Hymenoptera.

Finally I would like to thank the numerous people who have provided information, access to information, photographs and general advice:- Mr E. Brame, Mr R. Creevy, Dr P. Ellis, Mr S. J. Falk, Mr P. Ferris, Mrs P. Moxey (Epping Forest Conservation Centre), Mr Samarasingha (Epping Forest District Council) Mr D. Scott, Mr F. Scott, Mr C. Johnson, the curator and staff of the Passmore Edwards Museum, the staff of the reference section of Loughton Library and Mrs J. Bunyard who typed the manuscript.

Voucher Specimens

I have retained voucher specimens for all the invertebrate groups (except butterflies) and also for most mosses and among the higher plants for the grasses, sedges and rushes.

REFERENCES

1 ADDISON, W. (1977) *Portrait of Epping Forest.* Hale, London.
2 *Bacons New Map of Epping Forest, with Guide* (no date). London.
3 BAKER, C. A., MOXEY, P. A. & OXFORD, P. M. (1978) Woodland Continuity and Change in Epping Forest. *Field Studies* **4**: 645-669.
4 BAYNE, C. S. *et al* (1934) Birds of the London Area, *London Nat.* for 1933: 104 & 111.
5 BLOWER, J. G. (1958) British Millipedes (Diplopoda) Synopses of the British Fauna II, London (Linn. Soc.).
6 BOULGER, G. S. (1883) Epping Forest, its present condition, with suggestions for its future management; being reports resulting from recent visits to the Forest. *Trans. Essex Field Club* **3** pt. 7 Appendix No. 1: xxxix-xl.
7 BROOKS, P. T. & FULCHER, E. A. (1979) *An Archaeological Guide to Epping Forest.* Conservators of Epping Forest.
8 BUCK, F. D. (1955) A Provisional List of the Coleoptera of Epping Forest *Ent. Mon. Mag.* **91**: 174-192.
9 BUXTON, E. N. (1898) *Epping Forest* (5th Edition) Stanford, London.
10 CHAPMAN, D. I. (1977) Deer of Essex *Essex Nat.* (New Series) **1**: 43.
11 CHRISTY, R. M. (1890) *The Birds of Essex* Chelmsford.
12 COLE, H. A. (1887) Dormouse recorded from Buckhurst Hill. *Essex Nat.* **1**: 63.
13 COLE, W. (1894) The Management of Epping Forest. *Essex Nat.* **8**: 56.
14 CORKE, D. (Ed.) (1978) Epping Forest—the natural aspect? *Essex Nat.* (New Series) **2**.
15 CORKE, D. (Ed.) (1979) The Wildlife of Epping Forest. *Essex Nat.* (New Series) **4**.
16 CORNER, E. J. H. (1950) *A Monograph of Clavaria and Allied Genera.* Oxford.
17 DANIEL, W. B. (1812) *Rural Sports* **1**: 463. London.
18 DARLINGTON, A. (1975) *The Pocket Encyclopaedia of Plant Galls in Colour.* Revised Edition. Blandford.
19 DENNIS, R. W. G., ORTON, P. D. & HORA, F. B. (1960) New Checklist of British Agarics and Boleti. *Trans. Brit. Mycol. Soc. Suppl.*, 1-225.
20 EASON, E. H. (1964) *Centipedes of the British Isles.* Warne.
21 EDNEY, E. B. (1954) Woodlice. *Synopses of the British Fauna* **9**. London (Linn. Soc.).
22 ERITH, E. J. (no date) History of Chigwell Vols. 1 & 2. and other handwritten mss. in Loughton Library, includes records from the Rolls of the Court of Attachments of the Royal Forest of Waltham.
23 ERITH, E. J. (1950) Woodford, Essex 1600-1836. A Study of Local Government in a Residential Parish. Woodford and District Historical Society. *Proc. & Trans.* pt. **10** (see map facing p.124).
24 *Fairmead Lodge, Epping Forest* (1900) *Essex Nat.* **11**: 56.
25 *Field Meeting, July 1882. Proc. Essex Field Club* **2** pt. 6: xli.
26 FISHER, W. R. (1897) *The Forest of Essex.* Butterworth, London.
27 FORSYTH, L. (1978) Tiptree Heath—its history and natural history. *Essex Nat.* (New Series) **3**.
28 FREEMAN, J. (1862) *The Stratford Flora.* London.
29 GIBSON, G. S. (1862) *The Flora of Essex.* London.
30 GLEGG, W. E. (1929) *A History of the Birds of Essex.* London.
31 GRIGSON, G. (1975) *The Englishman's Flora.* Paladin.
32 HAWKSWORTH, D. L., JAMES, P. W. & COPPINS, B. J. (1980) Checklist of British Lichen-forming, Lichenicolous and Allied Fungi. *Lichenologist* **12**(1):1-115.
33 HUNTER, R. (1878) *The Epping Forest Act.* London.
34 JERMYN, S. T. (1975) *The Flora of Essex.* Essex Naturalists' Trust, Colchester.
35 JOHNSON, C. (1973) *Victorian Buckhurst Hill — a miscellany.* Epping Forest District council.
36 KERNEY, M. P. & CAMERON, R. A. D. (1979) *A Field Guide to the Land Snails of Britain and North West Europe.* Collins.
37 KLOET, G. S. & HINCKS, W. D. (1964) A Checklist of British Insects. Small orders & Hemiptera. *Handbk. Ident. Br. Insects* **11**(1)
38 KLOET, G. S. & HINCKS, W. D. (1972) A Checklist of British Insects. Lepidoptera. *Handbk. Ident. Br. Insects.* **11**(2).
39 KLOET, G. S. & HINCKS, W. D. (1977) A Checklist of British Insects. Coleoptera and Strepsiptera. *Handbk. Ident. Br. Insects.* **11**(3).
40 KLOET, G. S. & HINCKS, W. D. (1978) A Checklist of British Insects. Hymenoptera. *Handbk. Ident. Br. Insects.* **11**(4).
41 KLOET, G. S. & HINCKS, W. D. (1976) A Checklist of British Insects. Diptera & Siphonaptera. *Handbk. Ident. Br. Insects.* **11**(5).
42 LLOYD, E. G. (1977) The Wild Service tree *(Sorbus torminalis)* in Epping Forest. *London Nat.* **56**:22-28.
43 MALENOIR, G. (1963) Further Notes on the Reptile and Amphibian Survey of Epping Forest. *Essex Nat.* **31**: 142.
44 MORRIS, M. G. & PERRING, F. H. (Eds.)(1974) *The British Oak*, B.S.B.I., Classey.
45 MOSER, M. (1976) *Die Rohrlinge und Blatterpilze Kleine Kryptogamenflora.* Stuttgart.

46 *Ordnance Survey Sheet 177 East London* (revised to 1973). (Scale 1:50,000).
47 PARKER, E. J. (1974) Beech Bark Disease. Forestry Commission, *Forest Record* **96**.
48 PARMENTER, L. (1950) The Diptera of Bookham Common, *London Nat.* **29**: 98-133.
49 PARMENTER, L. (1960) A Further List of the Diptera of Bookham Common. *London Nat.* **39**: 66-76
50 PARMENTER, L. (1966) Some Additions to the List of Flies of Bookham Common. *London Nat.* **45**: 56-59.
51 PAULSON, R. (1900) An Inquiry into the causes of the Death of Birch Trees in Epping Forest and Elsewhere. *Essex Nat.* **11**: 273-284
52 PAYNE, R. M. (1968) The Crane-flies (Diptera: Tipulidae) of Epping Forest *Ent. Gaz.* **19**: 33-43.
53 PEGLER, D. N. (1967) Polyporaceae — Part II with a key to World Genera. *Bull. Brit. mycol. Soc.* **1**: 17-38.
54 PERCEVAL, P. J. S. (1909) *London's Forest*. Dent.
55 PETTIFER, A. J. (1968) A Bryophyte Flora of Essex. *Essex Nat.* **32**: 102.
56 PILAT, A. & USAK, O. (1958) *Mushrooms.*
57 *RACKHAM, O. (1975) Hayley Wood: its History and Ecology*. Cambs. and Isle of Ely Naturalists Trust Ltd., Cambridge.
58 RACKHAM, O. (1976) *Trees and Woodland in the British Landscape*. Dent.
59 RACKHAM, O. (1980) *Ancient Woodland, its history, vegetation and uses in England*. Arnold.
60 RAMSEY, A. R. J. (1953) Monkhams. *Woodford and District Historical Society* **11**: 1-12.
61 *Royal Institute of British Architects (West Essex Branch(*(1974) Buckhurst Hill A Planning Study.
62 SHENSTONE, J. C. (1894) The Oak Tree in Essex. *Essex Nat.* **8**: 98
63 STEELE, R. C. & WELCH, R. C. (1973) *Monks Wood. A Nature Reserve Record*. Nature Conservancy, Natural Environment Research Council.
64 STUBBS, A. E. & CHANDLER, P. J. (Eds.)(1978) *A Dipterist's Handbook*. Amateur Entomologists Society.
65 THOMPSON, P. (1921) On An Annotated Copy of Richard Warners Plantae Woodfordienses. *Essex Nat.* **19**: 81.
66 THOMPSON, P (1921) On Another Annotated Copy of Warners 'Plantae Woodfordienses'. *Essex Nat.* **19**: 234
67 THOMPSON, P. (1927) *Gnorimus nobilis* L. in Epping Forest. *Essex Nat.* **21**: 242.
68 *Tithe Award for Chigwell Manor 1838* (Loughton Library).
69 VERE, D. W. (1946) Hornets of Epping Forest *London Nat.* **26**: 99
70 *Victoria County History of Essex* Vol. 2 Industries: 448.
71 *Victoria County History of Essex* Vol. 4 Ongar Hundred. Chigwell: 18-43.
72 WALLER, W. C. (1905) The Foresters Walks in Waltham Forest. *Essex Review* **14**: 193-203.
73 WARD, B. (1979) *The Retreats of Epping Forest*. Conservators of Epping Forest.
74 WARNER, R. (1771) *Plantae Woodfordienses*. London.
75 WATLING, R. (1970) *The British Fungus Flora Vol. 1 Boletaceae.* H.M.S.O. (Edinburgh).

Deer of Essex

Essex Field Club
Passmore Edwards Museum,
Romford Road, Stratford, E15.

by Dr Donald Chapman

For fourteen years Donald Chapman ran the Essex Deer Survey in co-operation with the Essex Field Club and Essex Naturalists' Trust. The fascinating results of this most detailed survey of deer distribution in any British county are included in *Deer of Essex* — volume one of the important new *Essex Naturalist* series. Happily the survey shows that deer can co-exist with man; deer numbers are increasing and four species now breed in Essex compared with only one species fifty years ago.

Deer of Essex is a fifty page paperback, fully illustrated with photographs, maps and diagrams.
ISBN 0 905637 06 2 *Essex Naturalist No. 1* **£2.00**

Epping Forest - the natural aspect?

edited by Dr David Corke

In 1878 the Epping Forest Act charged the Forest Conservators with the preservation of the 'natural aspect of the Forest'. This book presents a summary of the soils and geological deposits on which the old Forest of Essex developed as the primaeval wildwood, and of how man's use of the Forest changed it over the centuries until it came under the management of the City of London. It also shows how the management decisions taken by the Forest Conservators have led to a serious loss of natural diversity. Recommendations are made as to how the Forest management could be improved.

Epping Forest — the natural aspect? is a 79 page paperback with many photographs figures and maps. Published 1978.
ISBN 0 905637 07 0 *Essex Naturalist No. 2* **£2.00**

Tiptree Heath — its history and natural history

by Laurie Forsyth

The author is organiser of the voluntary conservation group which has been largely responsible for the management of this important Essex Heath. An account of the prize winning conservation work is included, together with a fascinating account of the history and natural history of this site of special scientific interest.

Tiptree Heath — its history and natural history is a 19 page booklet with many photographs, drawings and a map. Published 1978.
ISBN 0 905653 08 9 *Essex Naturalist No. 3* **60p**

The Wildlife of Epping Forest

edited by Dr David Corke

A review of the changes in the Forest's animal life, especially over the first century of the Forest's management by the Corporation of the City of London. Based on a selection of the papers given at the Epping Forest Centenary Symposium, this book is an account of the Forest's land vertebrates (birds, mammals, reptiles and amphibians) and some of the most important groups of insects (butterflies, moths and beetles).

The Wildlife of Epping Forest is a 60 page paperback with seven plates and figures. Published 1979.

ISBN 0 905637 09 7 *Essex Naturalist No. 4* **£1.50**

A Short History – OF THE ESSEX FIELD CLUB 1880-1930

by Percy Thompson F.L.S.

Produced to celebrate the fiftieth anniversary of the Club, present day readers will be fascinated and amused by this account of the early activities of the Club. Partly based on the (sometimes acrimonious) debates recorded in the Club's Council minutes, this book also portrays the Club's presidents up to 1930.

A Short History of the Essex Field Club 1880-1930 is a 24 page paperback plus 4 pages of plates.

ISBN 0 905637 02 X **£1.00**

The Clay Tobacco-pipe in Britain

by Laurence S. Harley F.S.A.

Intended for the general reader with an interest in antiquities, Laurence Harley's book gives special attention to the history of clay tobacco-pipes in Essex and East Anglia. After an excellent introduction to the subject, enough detail is included to enable the reader to estimate the age of any pipe he finds and to discover something of the makers, from the marks on the pipes.

The Clay Tobacco-pipe in Britain is a fifty-one page paperback including nine pages of photographs and figures plus many text drawings. First published in 1963 it was revised and reprinted in 1976.

ISBN 0 905637 00 3 **£2.50**

Essex Field Club
Passmore Edwards Museum,
Romford Road, Stratford, E15.

FROM MEAN STREETS TO EPPING FOREST

THE SHAFTESBURY RETREAT LOUGHTON

BY
DAVID WILKINSON

LOUGHTON
THE LOUGHTON AND DISTRICT HISTORICAL SOCIETY
2000

ISBN 0952 88054 7

Published in 2000 by the Loughton and District Historical Society, and available from Forest Villa, Staples Road, Loughton, Essex, IG10 1HP.

Printed in Great Britain by
Riverside Press, Ipswich

Preface

In 1985 the Chigwell and Loughton History Society purchased a set of photographs that illustrated the outings organised by the Ragged School Union bringing children from the East End of London to Epping Forest between 1892 and 1939. As secretary of the Society at the time, I undertook a certain amount of research into the background of the outings, and the results of this, together with a selection of the photographs, form the substance of this book.

The number of children brought to Loughton was considerable – sometimes one thousand or more, six days a week, between June and September. Several of the older inhabitants of Loughton remember them well, and I have made much use of their memories. I have also received much help from Loughton Library (especially for local newspaper references and for back government records), from the Shaftesbury Society (as the Ragged School Union was renamed in 1914, whose records I have examined) and from colleagues in the Loughton and District Historical Society (formerly the Chigwell and Loughton History Society). Among these people, I pay special regard to the late Vic Barham, and I dedicate this book to his memory.

I have also received much help in the preparation of this book from Mrs Eve Lockington, Dr Donald Pohl, Dr Chris Pond, Mr Ted Martin and Mr Ian Strugnell, and I extend my warmest thanks to them.

In entitling the book *From Mean Streets to Epping Forest* I am also paying tribute to the Loughton author, Arthur Morrison, whose *Tales of Mean Streets* was published in 1901.

September 2000 DAVID WILKINSON

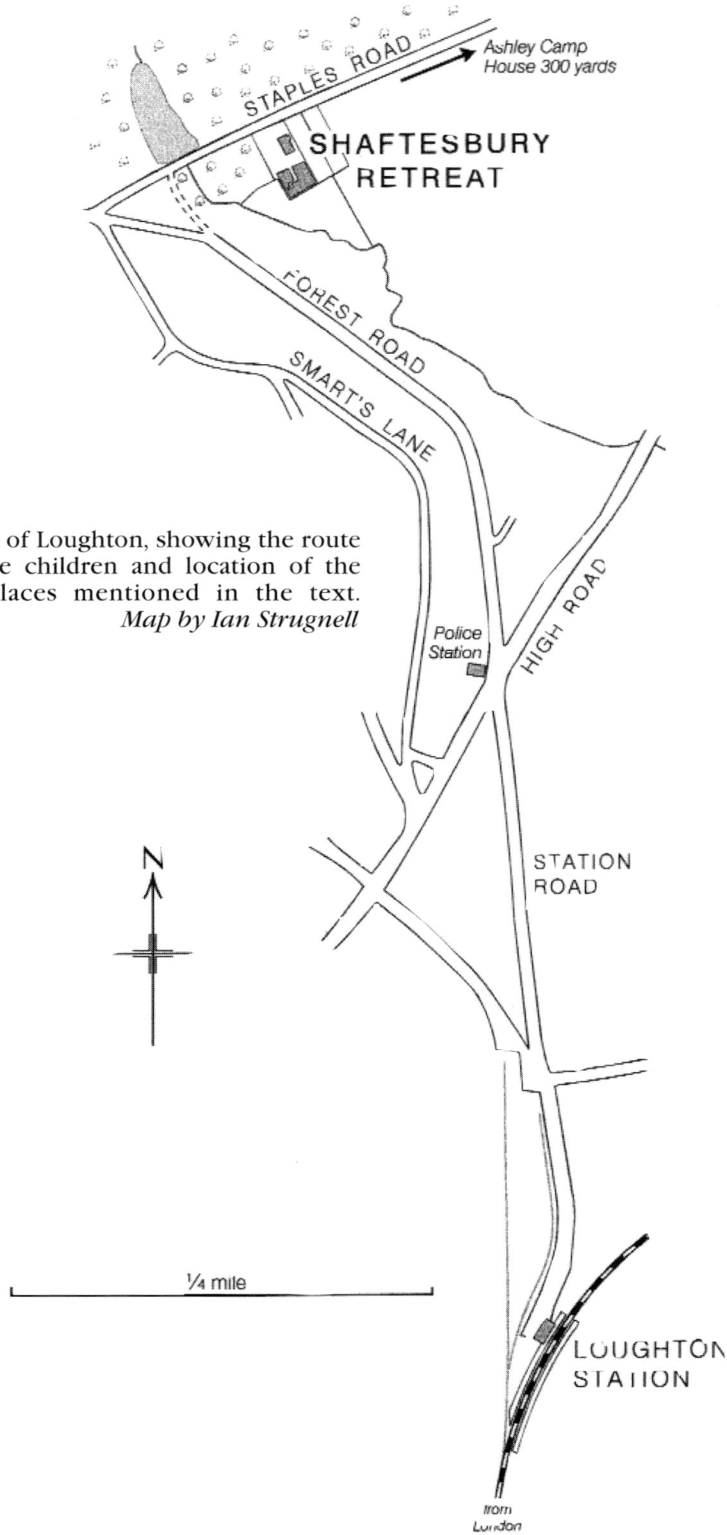

Map 1. Map of Loughton, showing the route taken by the children and location of the principal places mentioned in the text. *Map by Ian Strugnell*

Map 2. The Shaftesbury Retreat in Staples Road. *Map by Ian Strugnell*

From Mean Streets to Epping Forest

The idea of bringing children from London slums to the country originated with Cyril Arthur Pearson about 1890. At that time he was proprietor of *Pearson's Weekly*, and he was justifiably confident that he could raise the necessary finance through his journal. For organisation of the children's outings he turned to the Ragged School Union (RSU). Its secretary, John Kirk, devised the scheme whereby children could be taken to Epping Forest for a day at the cost of ninepence each. Pearson was delighted; ninepence was, in those days, the price of a cigar, and this suited his publicity projects admirably. He set up his Fresh Air Fund (FAF), with John Kirk and Ernest Kessel (his own secretary) as officers, and the scheme was put into practice.

Outings began in June 1892. In the first two years parties of children were taken from the East End by train to Snaresbrook and then marched to Wanstead Flats. But difficulties were encountered. Temporary accommodation had to be erected and dismantled daily, and members of the public proved troublesome. The RSU looked for property elsewhere, and found it in Loughton, at the Melbourne Retreat in Staples Road. It was owned by John Dean, a prominent local man, who offered refreshments to visitors to the Forest. He was willing to sell the property, and the RSU bought it. It was adapted to the new purpose and named the Shaftesbury Retreat (in honour of the founder of the RSU). A Mr A E Varney was appointed Supervisor, and RSU outings to Loughton began in July 1894.

The main problem of the early years at Loughton was the hostility of the local residents. The children were often dirty. Passengers complained of filthy trains, and sometimes, as the children walked through Loughton, a water cart had to follow them to cleanse the road. Their language was often obscene, and they were sometimes wantonly destructive. Some of the strong feelings aroused, appear from a unanimous resolution dated 1 April 1895 of the Loughton Parish Council:

> "That the clerk of the Parish Council, be instructed to write to the Secretary of the Ragged School Union, 27, Norfolk Street, Strand, and ask if it is the intention of that Society to repeat the practice of last July, August, and September by bringing down daily to the Shaftesbury Home, Staples Road, Loughton, some 500 or 600 children from the poorer districts of London. And if yes, would the Society please inform the Sanitary Committee of the Parish Council what steps will be taken this year by the Ragged School Union to prevent the poor children from loitering about the roads, obstructing the traffic, plundering the cottagers' gardens, littering the roads with paper and rubbish, defiling the Forest, bathing in the reservoir, and other indecent behaviour."

Over the years such feelings subsided. Mothers were encouraged to give their children a good wash before they set out. The Great Eastern Railway provided special trains for the visits and the RSU enforced some sort of discipline on the children. Local councillors were invited to visit the Shaftesbury Retreat and to observe these improvements. In 1912 the Urban District Council planned to re-lay a sewer across the grounds of the Retreat, and out of goodwill, the RSU waived any claim for compensation.

Regular routine soon established itself. A school in East

London closed for the day. All its children, accompanied by their teachers, came by train to Loughton. They then marched in procession via Station Road and Forest Road to the Retreat. There they were given meat pies and lemonade. Some children talked about frog pies, under the impression that country people ate frogs. Afterwards, they went across the road into the Forest. Games and races were organised; donkey rides and coconut shies were provided. Some children were given rides round the Forest in Ernest Kessel's car. Later in the afternoon they returned to the Retreat for tea. Then they returned to Loughton Station via Smarts Lane. This avoided upsetting local people in Forest Road twice in the same day. Sometimes the outward and inward routes were reversed, as in the cover picture.

There were some variations in the routine. Bad weather could spoil things. Accordingly clowns and other entertainers offered their services, and HMV donated a gramophone. Special arrangements had to be made for Jewish children with their dietary requirements. Disabled children (the contemporary word was cripples) needed special facilities – they were taken to Loughton by charabancs, and Boy Scouts were recruited from their home areas to assist with the transport. In September 1908 there was a special outing for grandmothers.

The outbreak of war in August 1914 cut down much of this activity, and although some parties did come to Loughton, efforts were made to organise outings elsewhere. Particular difficulty arose when the GER withdrew cheap fares. However outings were back to normal by 1919.

Pearson died in 1921. He had become a national figure when he bought out the *Daily Express* in 1900. But he suffered from glaucoma and went blind. He had been created a Baronet in 1916. He retained his enthusiasm for the FAF, and regularly visited the Retreat. His son, Sir Neville Pearson, proposed that in his father's honour, the first outing of the year, should take place on "Founder's Day" and that a distinguished visitor should be present. Sir Neville was himself married to the actress Gladys Cooper and so he was able to persuade several theatrical people to take their turn as visitors.

The Royal Family also took its turn in providing visitors. King George V, who had visited Loughton in 1908 as Prince of Wales, was greatly impressed, and regularly contributed to the FAF. He ensured that all his children came to Loughton – the Prince of Wales and Princess Mary, the Duke of York with his Duchess, the Duke of Gloucester and Prince George (later Duke of Kent). Prince George came in June 1931, and his success at the coconut shy was recorded in local newspapers as far afield as Aberdeen and Belfast.

The Shaftesbury Society (as the RSU had now renamed itself) began another project in Loughton. The former National School, at the other end of Staples Road, came on the market and was purchased. It was renamed Ashley Camp House (in memory of Lord Shaftesbury's family name), and was used for parties of children who would stay longer than a day. Parties came from the Society's missions in the country, and also from special groups sponsored by the LCC. At some period it was used by groups of mentally retarded children. However it did not fulfil all expectations. In May 1937 the property was sold for housing development, and Ashley Grove now marks the site.

With the outbreak of war in 1939, the regular outings ceased. The Chigwell Urban District Council took over the Shaftesbury Retreat, and converted it into an emergency mortuary for the expected victims of air raids. Even in 1947 the Shaftesbury Society hoped to resume pre-war outings, but conditions had changed, and the FAF found that the greater need was for longer holidays rather than for day trips. In October 1949, therefore, the Retreat was sold. For a while, the premises were used as the Loughton Belvedere

Glass Works, but housing development began in 1968. The association with the past has been preserved only in the name, Shaftesbury, for the new road that passes through its grounds. The Superintendent's House still stands at the corner of Staples Road and is clearly visible in the appropriate photographs.

The following is an extract from a letter written by Mrs I Alexander of Winchester, who in the 1930s lived at Canning Town:

"In the summer term there was talk of Loughton. We were blackmailed into good behaviour, running errands, saving our pennies and other inducements, otherwise the great day would be denied us. As I remember, the whole school, which I attended until 1936, was taken to Loughton. The day before, we had instructions on good behaviour. The night before, in our family at least, we were put in the zinc bath and scrubbed clean, and best clothes laid out. On the day itself we were given pocket money and went to school as usual. There, we were given a cardboard disc on string, to hang around our necks, and threatened with dire consequences should this be lost. I can remember it bore the words 'Ragged School Union' and 'Shaftesbury'. We couldn't think what it had to do with us. We were then marched to the station in crocodile fashion, along Bidder Street, ironically named after that great railway engineer and constructor of Victoria Docks. We sang as we marched, like soldiers on the move. 'It's a long way to Lousy Loughton' seemed a good song. We thought we had invented the term. Where was Tipperary, anyway? We were cheered the whole way. As we assembled on the station platform, people looked over the wall alongside Stephenson Street, named after that other great engineer and no fitting 'monument' to his memory. Stephenson Street was parallel with the railway line, and as the train moved out of the station, family and friends lining the street, waved us off, as teachers warned us in vain of the danger of hanging out of train windows. We continued to sing, through Stratford station and beyond until we arrived at Loughton station. We then walked to the 'big hall', which I now know to be the Shaftesbury Retreat. We probably had lunch then, and ate whatever was offered to us, before being let loose into the forest. We did the usual childish things – chased each other, fell in the mud and got lost. Getting lost was my greatest fear. One tree looked much like another. We were accustomed to using friends' houses as markers. At the end of the day we were all collected – I don't know how – and returned to the

'Retreat' for roll-call and tea. An annual ritual was the buying of gifts with our pocket money for every member of the family not lucky enough to be at Loughton. We bought such things as whistles, various 'blowers', brooches, cuff links, hair slides, balls, pencils and small ornaments for the mantelpiece. All were obtainable at the Retreat. It seemed a long day and the sun always shone. We all enjoyed ourselves and nobody ever forgot the experience. We were welcomed back and met at the station with smiling faces - or so it seemed."

Notes to the Photographs and Maps

Cover – Children walking from the Station to the Retreat at the Smarts Lane-Earls Path junction. Upper Belvedere Cottages, with billboards, in the background. This area of Forest, the Stubbles, is now overgrown with trees and bushes. The date is 1912.

Frontispiece – Sir Arthur Pearson

Map 1. Map of Loughton, showing the route taken by the children and location of the principal places mentioned in the text.

Map 2. The Shaftesbury Retreat in Staples Road.

Plate 1. Children mill around in the retreat yard. The hexagonal kiosk is where trinkets and souvenirs could be bought, behind which are the barn-like sheds where the children ate.

Plate 2. Having tea in one of the matchboarded sheds. Note the label worn by the boy on the right.

Plate 3. The gramophone donated by HMV was evidently used on fine as well as on wet days.

Plate 4. Staples Road is hardly changed today from this photograph of 1912. The Retreat House is just out of sight to the right. The two houses nearest the camera are Glendower, and then (with belvedere tower) Forest Villa, built by Robert Hunter in 1882 (on whom see separate LDHS monograph). The shed built in front of Nos 19-21 sold refreshments to ordinary visitors to the Forest. The children are being escorted by the FAF's uniformed janitor to their accustomed play area at Debden Slade, some ten minutes' walk into the Forest.

Plate 5. Staples Road Pond was commonly called "The Reservoir". But it was not a source of drinking water – it was built to retain floodwater. In the 1950s and 1960s, the pond became silted up, as a result of which the street known as Shaftesbury, built on the land of the Retreat, was severely flooded in 1982. The pond was redug, and a large earth dam constructed, in 1994-95: it will now retain the worst flash floods calculated to be expected in a 75-year period.

Plate 6. Games were played and races run in the Forest.

Plate 7. Clowns and donkeys were available to amuse the children. Note the railings in front of the Retreat House, which were saved from wartime salvage by Helen Sweeting, who occupied the retreat house from 1934 to her death in 2000.

Plate 8. Group photograph at Debden Slade.

Plate 9. Children lined up in front of the Retreat House, most likely in 1904. Note the much more open aspect of the forest abutting Staples Road in those days. The grassy banks have long been covered with quite mature trees. The sign describes "The Shaftesbury Holiday House" as the chief London centre of Pearson's Fresh Air Fund.

Plate 10. Staples Road enjoyed a Royal Visit almost every year, for which the locals would turn out in force. Joseph Hawkins Hayward, owner of Forest Villa (centre right), later patriotically had flagpole holders fixed to his house: the nine survive to this day. This scene shows the crowd waiting in Staples Road, probably in 1908. A police sergeant chats to bystanders whilst a cab is caught in the traffic.

Plate 11 (a) and (b). In 1923, the Royal visitors were TRH the Duke and Duchess of York, recently married, who of course later became King George VI and Queen Elizabeth.

Plate 12. In 1913, Lily Langtry, the actress, who whilst not Royalty, was well-known to Royalty, was the celebrity guest.

Plate 13. The Retreat House, renamed Melbourn Cottage, in the spring of 2000.

Plate 14. Children at the retreat with boxes of food from Gillard's, the Walthamstow pie and pickle makers.

Plate 15. Children buy trinkets and souvenirs to take home from the hexagonal kiosk provided in the yard of the Retreat.

Plate 1. Children mill around in the retreat yard.

Plate 2. Having tea in one of the matchboarded sheds.

Plate 3. The gramophone donated by HMV was evidently used on fine as well as on wet days.

Plate 4. Staples Road is hardly changed today from this photograph of 1912.

Plate 5. Staples Road Pond was commonly called "The Reservoir".

Plate 6. Games were played and races run in the Forest.

Plate 7. Clowns and donkeys were available to amuse the children.

Plate 8. Group photograph at Debden Slade.

Plate 9. Children lined up in front of the Retreat House, most likely in 1904.

Plate 10. Staples Road enjoyed a Royal Visit almost every year, for which the locals would turn out in force.

Plate 11 (a) and (b). In 1923, the Royal visitors were TRH the Duke and Duchess of York.

Plate 12. In 1913, Lily Langtry, the actress, was the celebrity guest.

Plate 13. The Retreat House, renamed Melbourn Cottage, in the spring of 2000.

Plate 14. Children at the retreat with boxes of food from Gillard's, the Walthamstow pie and pickle makers.

Plate 15. Children buy trinkets and souvenirs to take home.